SpringerBriefs in Cancer Research

More information about this series at http://www.springer.com/series/10786

Rajagopal N. Aravalli • Clifford J. Steer

Hepatocellular Carcinoma

Cellular and Molecular Mechanisms
and Novel Therapeutic Strategies

 Springer

Rajagopal N. Aravalli
Department of Radiology
University of Minnesota Medical School
Minneapolis, USA

Clifford J. Steer
Departments of Medicine and Genetics
  and Cell Biology and Development
University of Minnesota
Minneapolis, USA

ISBN 978-3-319-09413-7   ISBN 978-3-319-09414-4 (eBook)
DOI 10.1007/978-3-319-09414-4
Springer Cham Heidelberg New York Dordrecht London

Library of Congress Control Number: 2014946590

Printed on acid-free paper

Springer is part of Springer Science+Business Media (www.springer.com)

Contents

1 Introduction 1
References 2

2 Etiology 3
Risk Factors 3
Demographics 4
References 5

3 Current Diagnosis and Treatment Options for HCC 7
Surgical Interventions 7
Percutaneous Interventions 8
Transarterial Interventions 9
Drug Treatment 9
Challenges for Treatment 9
   Genetic Heterogeneity 10
   Inadequate Treatment for Tumor Local Recurrence 11
   Accurate Stratification of Patient Risk with Underlying Liver Disease ... 11
References 12

4 Pathophysiology of HCC 15
Tumor Microenvironment 16
   Cancer-Associated Fibroblasts 16
   Tumor-Associated Macrophages 16
   Dendritic Cells 17
   Hepatic Stellate Cells 17
   Endothelial Cells 17
   T Cells 18
   The Extracellular Matrix 18
Hypoxia 19
Epithelial-Mesenchymal Transition 21
Oxidative Stress 21

Cancer Stem Cells 22
Inflammation 24
References 26

5 Molecular Mechanisms of HCC 33
Signaling Pathways 33
    Wnt/β-Catenin Signaling Pathway 34
    p53 Pathway 35
    Rb Pathway 35
    Ras Pathway 36
    MAPK Pathway 36
    JAK/STAT Pathway 37
    Heat Shock Proteins 38
    Others 38
MicroRNAs 39
References 41

6 Animal Models of Liver Cancer 47
Mouse 47
Rat 48
Rabbit 48
Pig 48
Primate 49
Woodchuck 49
A Case for Better Animal Models 50
References 50

7 Novel Therapeutic Strategies to Combat HCC 51
Small Molecule-Based Therapeutics 51
MicroRNA Therapeutics 55
    Inhibition of OncomiRs 55
    MicroRNA Replacement Therapy 57
Stem Cell Therapy 59
References 60

8 Conclusions 65
References 66

Index 67

Chapter 1
Introduction

In general, primary liver cancers can be categorized as hepatoblastoma, hepatocellular carcinoma (HCC), angiosarcoma, and cholangiocarcinoma. The most common of these is HCC, which accounts for nearly 80 % of all liver cancer cases. HCC is the sixth most common cancer, and third leading cause of cancer-related deaths worldwide (Nordenstedt et al. 2010). It normally develops as a consequence of underlying chronic liver disease and portal hypertension, and is frequently associated with cirrhosis. Although HCC is diagnosed at a late stage in most cases, current treatment options to treat cancer patients are limited to surgical resection and whole or partial liver transplantation. Because liver resection is only offered in a few cases due to the high morbidity and mortality in patients with cirrhosis and portal hypertension, orthotopic liver transplantation (OLT) is the most common option in patients with end-stage liver disease and HCC (Hernandez-Gea et al. 2013). Recurrence or metastasis is quite common in patients who have had a resection, and survival rate is 30–40 % at 5 years post-surgery (Aravalli et al. 2008). While the recurrence rate is between 8 and 15 % in HCC individuals after OLT, this rate is much higher after surgical resection and ablation therapy (Mazzaferro et al. 1996).

In addition to human loss, there is an enormous economic impact of HCC. Using data from the Surveillance, Epidemiology and End Results (SEER) and Medicare dataset of elderly patients, it was recently reported that the annual healthcare costs associated with HCC in the United States alone, where it accounts only for 2.9 % all cancer-related deaths, were estimated to be about $455 million (Lang et al. 2009). Typically, overall costs are approximately 6- to 8-fold higher than patients with liver disease but without HCC, underscoring the substantial burden of medical care expenses associated with HCC (White et al. 2012). As the incidence of HCC is expected to increase in the next decade, due mainly to viral hepatitis, alcoholism, and non-alcoholic steatohepatitis, it is crucial that we develop more effective diagnostic, predictive and therapeutic options.

© The Author(s) 2014
R.N. Aravalli, C.J. Steer, *Hepatocellular Carcinoma*, SpringerBriefs
in Cancer Research, DOI 10.1007/978-3-319-09414-4_1

References

Aravalli RN, Steer CJ, Cressman EN (2008) Molecular mechanisms of hepatocellular carcinoma. Hepatology 48(6):1049–1053

Hernandez-Gea V, Turon F, Berzigotti A, Villanueva A (2013) Management of small hepatocellular carcinoma in cirrhosis: focus on portal hypertension. World J Gastroenterol 19(8):1193–1199

Lang K, Danchenko N, Gondek K, Shah S, Thompson D (2009) The burden of illness associated with hepatocellular carcinoma in the United States. J Hepatol 50(1):89–99

Mazzaferro V, Regalia E, Doci R, Andreola S, Pulvirenti A, Bozzetti F, Montalto F, Ammatuna M, Morabito A, Gennari L (1996) Liver transplantation for the treatment of small hepatocellular carcinomas in patients with cirrhosis. N Eng J Med 334(11):693–700

Nordenstedt H, White DL, El-Serag HB (2010) The changing pattern of epidemiology in hepatocellular carcinoma. Dig Liver Dis 42(Suppl 3):S206–S214

White LA, Menzin J, Korn JR, Friedman M, Lang K, Ray S (2012) Medical care costs and survival associated with hepatocellular carcinoma among the elderly. Clin Gastroenterol Hepatol 10(5):547–554

Chapter 2
Etiology

HCC is a heterogeneous cancer caused by a variety of risk factors including alcohol and more recently, the metabolic syndrome (Aravalli et al. 2008). The myriad of factors varies according to the geographical region, thereby complicating the diagnosis, prognosis and treatment recommendations (Marrero et al. 2010). Even though most of the 600,000 new HCC cases that occur each year are from developing countries, the incidence of HCC is rising rapidly in Western countries. This is due in large part to hepatitis C, alcoholism, and obesity (Fletcher and Powell 2003; El-Serag 2012). Although screening blood for HCV began in the 1990s, those patients infected beforehand are now presenting with HCC and therefore, many individuals remain undiagnosed. The Centers for Disease Control recently recommended one-time screening for the entire generation born between 1945 and 1965. The growing incidence of HCC has generated intense research efforts to understand physiological, cellular, and molecular mechanisms of the disease with the hope of developing new treatment strategies.

Risk Factors

A variety of factors have been firmly identified in the development of liver cancer (Fig. 2.1). Such factors include infection with viruses (Cramp 1999; Blum 2005; El-Serag and Rudolph 2007), exposure to foodstuffs contaminated with aflatoxin B1 (AFB1) (Soini et al. 1996), and vinyl chloride (Boffetta et al. 2003), tobacco (Tsukuma et al. 1995), heavy alcohol intake (Donato et al. 2002), non-alcoholic fatty liver disease (Hashimoto et al. 2004), diabetes (Wideroff et al. 1997; Regimbeau et al. 2004), obesity (Regimbeau et al. 2004), diet (Yu et al. 1995), coffee (Kurozawa et al. 2005), oral contraceptives (Maheshwari et al. 2007), and hemochromatosis, to name a few (Hellerbrand et al. 2003). In general, these factors vary according to the geographical region. For instance, chronic hepatitis B virus (HBV) infection

© The Author(s) 2014
R.N. Aravalli, C.J. Steer, *Hepatocellular Carcinoma*, SpringerBriefs
in Cancer Research, DOI 10.1007/978-3-319-09414-4_2

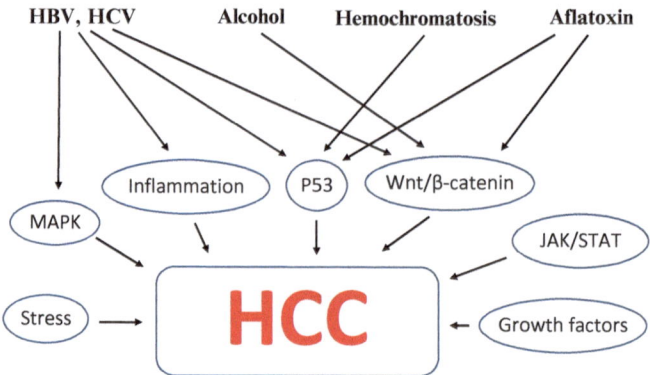

Fig. 2.1 Cellular signaling pathways known to be activated by various risk factors to induce HCC

is common in many countries in Asia and sub-Saharan Africa, whereas HCV is prevalent in Japan, Western Europe, and the United States (El-Serag and Rudolph 2007). Such differences in infectivity add complexity to the extrapolation of data obtained from one geographical region and applying it to others.

Demographics

The prognosis for liver cancer is very poor and the geographical patterns of incidence and mortality are similar. Analysis of the data compiled by the International Agency for Research on Cancer (IARC) has shown that 83 % of the estimated 782,000 new liver cancer cases that occurred in 2012 were from developing countries, with China alone accounting for 50 % (IARC 2012). While liver cancer is the fifth most common cancer in men and the ninth in women, it is the second most common cause of death from cancer worldwide (IARC 2012). Due to the prevalence of infection with hepatitis viruses, HCC has become one of the top three cancers in the Asia-Pacific region. Whereas HBV is predominant in most Asian countries, HCV is far more prevalent in Japan, Australia, and New Zealand (Yuen et al. 2009).

Age-standardized incidence rates, have shown that HCC occurrence worldwide is considerably higher in men than in women (Nordenstedt et al. 2010). The incidence of HCC in Asian patients was found to be 31 % in males and 18 % in females, while there was no difference in patients with HCV (Miyakawa et al. 1996). HCC among HBV carriers in Asia exceeds 0.2 % by the age of 40; and therefore, surveillance was recommended in males older than 40 years and females older than age 50, even in the absence of cirrhosis (Yu et al. 2000). In a cohort of 391 HCC patients from South Africa, serum levels of HBV were significantly higher in 171 out of 173 individuals below the age of 30, potentially due to aflatoxin exposure and synergy with HBV, prompting the authors to suggest surveillance at any early age (Kew and

Macerollo 1988). Interestingly, HBV-infected individuals exposed to aflatoxin are 60 times more likely to develop HCC than those with no exposure (Qian et al. 1994). As shown in a recent study from Taiwan, heavy alcohol consumption significantly increased the incidence of HCC in patients with HBV-associated cirrhosis (Lin et al. 2013). In North America and Europe, HCV is the major etiological factor of HCC. However, HCV infection is primarily limited to patients with cirrhosis (Fattovich et al. 1997), and cirrhosis amplifies the risk of HCC among patients with chronic viral hepatitis (Kuper et al. 2001). In these geographical locations, alcoholic cirrhosis and nonalcoholic steatohepatitis (NASH) are also major risk factors for HCC development. Diabetes mellitus may also increase the risk of developing liver cancer (Adami et al. 1996). This was highlighted in a recent study with US veterans, where HCC was found to be overwhelmingly associated with HBV, HCV and alcohol consumption, as well as diabetes with no evidence of cirrhosis (Karagozian et al. 2013). Collectively, these studies demonstrated key differences in age, gender, and the nature and synergy of risk factors, and underscore the importance of early detection for the treatment of HCC.

References

Adami HO, Chow WH, Nyrén O, Berne C, Linet MS, Ekbom A, Wolk A, McLaughlin JK, Fraumeni JF Jr (1996) Excess risk of primary liver cancer in patients with diabetes mellitus. J Natl Cancer Inst 88(20):1472–1477

Aravalli RN, Steer CJ, Cressman EN (2008) Molecular mechanisms of hepatocellular carcinoma. Hepatology 48(6):1049–1053

Blum HE (2005) Hepatocellular carcinoma: therapy and prevention. World J Gastroenterol 11(47):7391–7400

Boffetta P, Matisane L, Mundt KA, Dell LD (2003) Meta-analysis of studies of occupational exposure to vinyl chloride in relation to cancer mortality. Scand J Work Environ Health 29(3):220–229

Cramp M (1999) HBV + HCV = HCC? Gut 45(2):168–169

Donato F, Tagger A, Gelatti U, Parrinello G, Boffetta P, Albertini A, Decarli A, Trevisi P, Ribero ML, Martelli C, Porru S, Nardi G (2002) Alcohol and hepatocellular carcinoma: the effect of lifetime intake and hepatitis virus infections in men and women. Am J Epidemiol 155(4):323–333

El-Serag HB (2012) Epidemiology of viral hepatitis and hepatocellular carcinoma. Gastroenterology 142(6):1264–1273

El-Serag HB, Rudolph KL (2007) Hepatocellular carcinoma: epidemiology and molecular carcinogenesis. Gastroenterology 132(7):2557–2576

Fattovich G, Giustina G, Degos F, Tremolada F, Diodati G, Almasio P, Nevens F, Solinas A, Mura D, Brouwer JT, Thomas H, Njapoum C, Casarin C, Bonetti P, Fuschi P, Basho J, Tocco A, Bhalla A, Galassini R, Noventa F, Schalm SW, Realdi G (1997) Morbidity and mortality in compensated cirrhosis type C: A retrospective follow-up study of 384 patients. Gastroenterology 112(2):463–472

Fletcher LM, Powell LW (2003) Hemochromatosis and alcoholic liver disease. Alcohol 30(2):131–136

Hashimoto E, Taniai M, Kaneda H, Tokushige K, Hasegawa K, Okuda H, Shiratori K, Takasaki K (2004) Comparison of hepatocellular carcinoma patients with alcoholic liver disease and nonalcoholic steatohepatitis. Alcohol Clin Exp Res 28(8 Suppl Proceedings):164S–168S

Hellerbrand C, Pöppl A, Hartmann A, Schölmerich J, Lock G (2003) HFE C282Y heterozygosity in hepatocellular carcinoma: evidence for an increased prevalence. Clin Gastroenterol Hepatol 1(4):279–284

IARC (2012) GLOBOCAN 2012: estimated cancer incidence, mortality and prevalence worldwide in 2012. International Agency for Research on Cancer (IARC). http://globocan.iarc.fr/Default. aspx

Karagozian R, Baker E, Houranieh A, Leavitt D, Baffy G (2013) Risk profile of hepatocellular carcinoma reveals dichotomy among US veterans. J Gastrointest Cancer 44(3):318–324

Kew MC, Macerollo P (1988) Effect of age on the etiologic role of the hepatitis B virus in hepatocellular carcinoma in blacks. Gastroenterology 94(2):439–442

Kuper H, Ye W, Broomé U, Romelsjö A, Mucci LA, Ekbom A, Adami HO, Trichopoulos D, Nyrén O (2001) The risk of liver and bile duct cancer in patients with chronic viral hepatitis, alcoholism, or cirrhosis. Hepatology 34(4 Pt 1):714–718

Kurozawa Y, Ogimoto I, Shibata A, Nose T, Yoshimura T, Suzuki H, Sakata R, Fujita Y, Ichikawa S, Iwai N, Tamakoshi A, JACC Study Group (2005) Coffee and risk of death from hepatocellular carcinoma in a large cohort study in Japan. Br J Cancer 93(5):607–610

Lin CW, Lin CC, Mo LR, Chang CY, Perng DS, Hsu CC, Lo GH, Chen YS, Yen YC, Hu JT, Yu ML, Lee PH, Lin JT, Yang SS (2013) Heavy alcohol consumption increases the incidence of hepatocellular carcinoma in hepatitis B virus-related cirrhosis. J Hepatol 58(4):730–735

Maheshwari S, Sarraj A, Kramer J, El-Serag HB (2007) Oral contraception and the risk of hepatocellular carcinoma. J Hepatol 47(4):506–513

Marrero JA, Kudo M, Bronowicki JP (2010) The challenge of prognosis and staging for hepatocellular carcinoma. Oncologist 15(suppl 4):23–33

Miyakawa H, Izumi N, Marumo F, Sato C (1996) Roles of alcohol, hepatitis virus infection, and gender in the development of hepatocellular carcinoma in patients with liver cirrhosis. Alcohol Clin Exp Res 20(1 Suppl):91A–94A

Nordenstedt H, White DL, El-Serag HB (2010) The changing pattern of epidemiology in hepatocellular carcinoma. Dig Liver Dis 42(Suppl 3):S206–S214

Qian GS, Ross RK, Yu MC, Yuan JM, Gao YT, Henderson BE, Wogan GN, Groopman JD (1994) A follow-up study of urinary markers of aflatoxin exposure and liver cancer risk in Shanghai, People's Republic of China. Cancer Epidemiol Biomarkers Prev 3(1):3–10

Regimbeau J, Colombat M, Mognol P, Durand F, Abdalla E, Degott C, Degos F, Farges O, Belghiti J (2004) Obesity and diabetes as a risk factor for hepatocellular carcinoma. Liver Transplant 10(2 Suppl 1):S69–S73

Soini Y, Chia SC, Bennett WP, Groopman JD, Wang JS, DeBenedetti VM, Cawley H, Welsh JA, Hansen C, Bergasa NV, Jones EA, DiBisceglie AM, Trivers GE, Sandoval CA, Calderon IE, Munoz Espinosa LE, Harris CC (1996) An aflatoxin-associated mutational hotspot at codon 249 in the p53 tumor suppressor gene occurs in hepatocellular carcinomas from Mexico. Carcinogenesis 17(5):1007–1012

Tsukuma H, Hiyama T, Oshima A, Sobue T, Fujimoto I, Kasugai H, Kojima J, Sasaki Y, Imaoka S, Horiuchi N et al (1995) A case–control study of hepatocellular carcinoma in Osaka, Japan. Int J Cancer 45(2):231–236

Wideroff L, Gridley G, Mellemkjaer L, Chow WH, Linet M, Keehn S, Borch-Johnsen K, Olsen JH (1997) Cancer incidence in a population-based cohort of patients hospitalized with diabetes mellitus in Denmark. J Natl Cancer Inst 89(18):1360–1365

Yu MW, Hsieh HH, Pan WH, Yang CS, Chen CJ (1995) Vegetable consumption, serum retinol level, and risk of hepatocellular carcinoma. Cancer Res 55(6):1301–1305

Yu MW, Chang HC, Liaw YF, Lin SM, Lee SD, Liu CJ, Chen PJ, Hsiao TJ, Lee PH, Chen CJ (2000) Familial risk of hepatocellular carcinoma among chronic hepatitis B carriers and their relatives. J Natl Cancer Inst 92(14):1159–1164

Yuen MF, Hou JL, Chutaputti A (2009) Hepatocellular carcinoma in the Asia pacific region. J Gastroenterol Hepatol 24(3):346–353

Chapter 3
Current Diagnosis and Treatment Options for HCC

The clinical features of HCC may be nonspecific and difficult to diagnose. They include weight loss, deteriorating liver function in patients with cirrhosis, and in the later stages, the tumor may result in significant abdominal pain in the right upper quadrant. In rare, unfortunate cases intra-abdominal bleeding due to rupture of the liver tumor can occur; and this is associated with a high rate of mortality. Screening of at-risk patients is thus an important factor in early diagnosis and treatment. This is underscored by a large cohort study of 461 Italian patients with HCC in which a significant proportion (23 %) were asymptomatic (Trevisani et al. 1996). In recent years, it has becoming increasingly common that asymptomatic patients were being diagnosed for HCC during routine screening for cirrhosis or as a candidate for liver transplantation (Wurmbach et al. 2007). Cirrhosis from any cause predisposes to HCC, and unfortunately, the disease is quite advanced at this stage and therapeutic options are limited. Even though many screening protocols have been suggested to balance costs and accuracy, there is still an unmet need given how quickly the tumors tend to grow. The difficulty lies in finding a test with adequate sensitivity and specificity to identify patients while they are still candidates for successful resection. During the past decade, advances in imaging techniques such as magnetic resonance imaging (MRI) and contrast-enhanced ultrasonography with real-time low mechanical index harmonic imaging ultrasound equipment have allowed us to perform non-invasive diagnosis of HCC. They also replaced the need in many cases for invasive procedures like ultrasound-guided biopsy or angiography (Piscaglia and Bolondi 2007).

Surgical Interventions

For patients who are fortunate enough to be identified while still candidates for resection, surgery is potentially curative. Whole liver transplant is the ideal option as it treats both the tumor and the underlying cirrhosis. However, waiting lists are

© The Author(s) 2014
R.N. Aravalli, C.J. Steer, *Hepatocellular Carcinoma*, SpringerBriefs
in Cancer Research, DOI 10.1007/978-3-319-09414-4_3

notably long and patients either die or tumor progression leads to disqualification during the waiting period. A separate issue outside the scope of this review is the selection of patients for liver transplantation. The shortage of donor livers has led to the development of techniques such as split liver transplantation and living donor transplantation (Gruessner 2008). It is difficult to assess hepatic functional reserve but this is a key issue regarding suitability for surgery beyond the usual concerns of size, multiplicity, and location (Ribero et al. 2008). It must also be borne in mind that the remnant liver of course is still severely diseased with the same history as the tumorous tissue. Unrecognized micrometastatic disease, local recurrence, and *de novo* tumor development elsewhere in the liver occur frequently (Taura et al. 2006). There is ongoing debate as to the merits of various resection techniques such as enucleation, segmentectomy, and lobectomy, depending upon the circumstances. A procedure that has been used preoperatively around the world and, more recently at several centers in the United States, is portal vein embolization (de Baere et al. 2007). In this technique, the involved lobe is embolized via access to the portal vein, thereby causing atrophy of the affected side through deprivation of trophic factors and nutrient flow. Resection is then performed several weeks later after the remnant liver has had some time to hypertrophy and establish normal liver function.

Percutaneous Interventions

A number of image-guided interventions have been developed over the years (Garcea et al. 2003). Initially, tumor ablation was performed by injection of ethanol and later by use of 50 % acetic acid (Clark and Soulen 2005). These methods are effective but require multiple procedures to effectively treat both tumor and surrounding tissue. The ratio of injection volume to desired ablation volume is about 1:1 for ethanol and about 1/3 of that for acetic acid. The volume of ethanol that most operators are comfortable injecting is about 10 ml. Although acetic acid has been shown to be a more effective agent, the safe injection volume is in the range of 3 ml. Thus the same number of procedures is required regardless of the agent.

A number of additional concerns such as tract reflux and irregular distribution along paths of least resistance have led to the development of other methods such as hyperthermia and cryotherapy. Hyperthermia can be delivered by several techniques, including radiofrequency ablation, microwave, laser therapy, and less commonly high-intensity focused ultrasound (Gannon and Curley 2005). Effective cryotherapies are focused on developing better delivery of gas for cooling and in decreasing the size of probes (Hinshaw and Lee 2007). Thermal therapies, whether hot or cold, are associated with troublesome heat sink issues. As an example, tissue adjacent to larger blood vessels becomes extremely difficult to treat due to the dissipation of thermal effects in these areas much like a radiator in a car. Single use devices, anesthesia, and the capital required for base units such as power generators add significantly to the cost of a procedure but small tumors often may be treated with a single ablation.

Transarterial Interventions

While data are beginning to accrue for the ablation methods and survival benefits, studies have shown a survival benefit from catheter-based techniques (Llovet et al. 2002). As always, careful patient selection remains a key issue in achieving successful treatment. Although initially, simple arterial infusion of drugs was the standard, increased survival was achieved using embolic particles to induce ischemia and prolong dwell times of drugs via increased local concentrations with relatively minimal systemic effect. In the past decade the embolic agents themselves have become the subject of much research, including the use of yttrium-90 for example, drug-eluting beads, or beads made of various materials that respond in different ways to the environment both in terms of deformability and degradation over time. There is even debate over the need for any associated drug in light of the ischemia produced by the injected particles (Covey et al. 2006).

Drug Treatment

In recent years, a number of studies have employed drug treatments to target cellular signaling pathways. These include small-molecule protein kinase inhibitors, monoclonal antibodies, and antibiotics either alone or in combinations (Skelton and O'Neill 2008). One such study examined whether interferon therapy prevents the development of HCC in patients with chronic HCV infection (Okanoue et al. 1999). Following therapy, the patients were studied between 1 to 7 years by performing blood tests and image analysis to detect HCC. The cumulative incidence of HCC was found to be significantly lower in cirrhotic patients, but not in patients in advanced stages. Similar results were reported in chronic HCV patients undergoing IFN therapy (Ikeda et al. 1999). More recently, sorafenib showed a modest, 3-month survival benefit in selected patients (Skelton and O'Neill 2008). Prior to this, however, over 100 clinical trials of IV chemotherapy using many drugs or combinations for HCC showed minimal response to treatment and failed to demonstrate any statistically significant survival benefit (Thomas et al. 2008).

Challenges for Treatment

Screening and surveillance is a common practice in the clinical management of patients with chronic liver disease. Follow up schedules may vary according to clinic and/or physician's policies, and intervals of about 6 months are typical. At present, the major challenge is to detect, confirm, and stage HCC at an early stage. It was proposed that only individuals diagnosed with HCC should be entered into surveillance, and should be stopped if severe associated conditions or liver failure precluding treatment appear during screening (Bolondi 2003). Moreover, chronic hepatitis and cirrhosis frequently co-exist at an early state and differential diagnosis between the two conditions is difficult with current diagnostic methods, including biopsy (Gaiani et al. 1997; Kudo et al. 2008).

The presence and detection of a nodule in a cirrhotic liver during a screening does not necessarily correlate with a positive diagnosis for HCC. However, after detection, a series of diagnostic and interventional procedures should be followed in the clinic to validate the presence of tumor(s). The rate and the subsequent management of suspected nodules was not adequately reported in the literature suggesting that proper assessment of tumor nodules is a persistent problem (Bolondi 2003). The incidence of nodules in which a definitive diagnosis of cancer cannot be made was estimated to be as high as 17 %, and this was due to the presence of dysplastic nodules or other benign and malignant conditions which may arise in cirrhotic livers (Bolondi 2003). Major challenges that we face in the clinical management of HCC are:

1. genetic heterogeneity
2. treating tumor local recurrence adequately
3. minimizing damage to non-tumor tissue to preserve the already diseased liver
4. accurately stratifying the patient for underlying liver disease and risk
5. recurrent disease elsewhere in the liver – sometimes this might even be acceler- ated due to treatment in another location

Genetic Heterogeneity

One of the major challenges in the management of HCC is the existence of a patient population with significant genetic heterogeneity. Even though various environmental risk factors do play a role in the risk of HCC, individual genetic predisposition can also cause HCC. This was underscored by the fact that about 20 % of cases diagnosed in the United States were not associated with any risk factors, including viral hepatitis and alcohol (El-Serag and Mason 2000). A number of chromosomal loci affecting genetic susceptibility to chemically induced liver cancer was identified in mice and rats (Dragani et al. 1996; Feo et al. 2006), strongly suggesting that genetic heterogeneity plays a role in hepatocarcinogenesis.

At present a 'one-size-fits-all approach' of transarterial chemoembolization (TACE) administration is being adapted in the clinic, and 'unifocal' nodules are subjected to ablation. Biopsies are no longer carried out routinely because of the American Association for the Study of Liver Diseases (AASLD) criteria that stipulates them only when AFP > 400 and imaging (CT or MRI) is >2 cm. In fact, a recent survey showed that Canadian physicians do not perform routine assessment of liver fibrosis (Sebastiani et al. 2014). Nowadays, most physicians employ noninvasive methods, such as FibroScan, even though limitations in access to and availability of noninvasive methods is a significant barrier (Stevenson et al.

2012; Aljawad et al. 2013; Sebastiani et al. 2014). However, these methods do not eliminate the concern of genetic heterogeneity which is more evident in the intermediate stages of HCC than in the early or advanced stages of the disease (Bolondi et al. 2012). Therefore, there is a greater need to tailor the treatments based on the evidence obtained from patient samples. Towards this, an expert panel recently proposed a set of guidelines that contain four substages of intermediate stage in HCC patients that will facilitate better management of disease (Bolondi et al. 2012).

Inadequate Treatment for Tumor Local Recurrence

Surgical resection remains the established modality for HCC (Wong and Lo 2013; Li et al. 2014). While resection has improved survival rates, overall survival rates post-resection in patients with multiple HCCs are still abysmal and recurrence is quite common in patients who have had resection (Bruix and Sherman 2005; Llovet et al. 2012). Inadequate treatment for tumor recurrence is therefore a significant problem in reducing the mortality in patients diagnosed with HCC.

A meta-analysis of nine anatomic and non-anatomic resection studies recently reported that patients with anatomic resection (AR) had a longer survival rate than non-AR (NAR) (Chen et al. 2011). However, caution should still be exercised in the interpretation of these results because of genetic heterogeneity in patients. A recent study with data obtained from 543 cirrhotic patients of Child-Pugh class A that were submitted either to AR or NAR showed that while AR for early HCC could result in lower early recurrence of tumors, NAR does not increase the recurrence rate in differentiated tumors or in the absence of microvascular invasion (Cucchetti et al. 2014).

Accurate Stratification of Patient Risk with Underlying Liver Disease

Targeted screening strategies based on stratification of patients according to risk is an attractive and logical approach for health economics. However, these are limited by the number and variability of risk factors, making it difficult to provide an effective individual risk score that can be applied to all geographical areas equally (Bolondi 2003). Moreover, older age and advanced cirrhosis with functional impairment of the liver are hindrances that reduce the chance of HCC patients to benefit from curative treatments. Different screening schedules based on risk factors are difficult to conceive, and therefore any change to the current diagnostic tests (ultrasonography and AFP) is not forthcoming. Liver transplantation remains the most effective viable treatment option, and the number of patients receiving this treatment is increasing steadily due to the introduction of the Model for End-stage

Liver Disease (MELD) score (Washburn 2010), and the improvement in screening criteria for liver transplantation will continue to be a daunting task in the years ahead.

References

Aljawad M, Yoshida EM, Uhanova J, Marotta P, Chandok N (2013) Percutaneous liver biopsy patterns among Canadian hepatologists. Can J Gastroenterol 27(11):e31–e34

Bolondi L (2003) Screening for hepatocellular carcinoma in cirrhosis. J Hepatol 39(6):1076–1084

Bolondi L, Burroughs A, Dufour JF, Galle PR, Mazzaferro V, Piscaglia F, Raoul JL, Sangro B (2012) Heterogeneity of patients with intermediate (BCLC B) hepatocellular carcinoma: proposal for a subclassification to facilitate treatment decisions. Semin Liver Dis 32(4): 348–359

Bruix J, Sherman M, Practice Guidelines Committee, American Association for the Study of Liver Diseases (2005) Management of hepatocellular carcinoma. Hepatology 42(5):1208–1236

Chen J, Huang K, Wu J, Zhu H, Shi Y, Wang Y, Zhao G (2011) Survival after anatomic resection versus nonanatomic resection for hepatocellular carcinoma: a meta-analysis. Dig Dis Sci 56(6):1626–1633

Clark T, Soulen MC (2005) Chemical ablation of hepatocellular carcinoma. J Vasc Interv Radiol 13(9 Pt2):S245–S252

Covey AM, Maluccio MA, Schubert J, BenPorat L, Bordy LA, Sofocleous ST, Getrajdman GI, Fong Y, Brown KT (2006) Particle embolization of recurrent hepatocellular carcinoma after hepatectomy. Cancer 106(10):2181–2189

Cucchetti A, Qiao GL, Cescon M, Li J, Xia Y, Ercolani G, Shen F, Pinna AD (2014) Anatomic versus nonanatomic resection in cirrhotic patients with early hepatocellular carcinoma. Surgery 155(3):512–521

de Baere T, Denys A, Madoff DC (2007) Preoperative portal vein embolization: indications and technical consideration. Tech Vasc Interv Rad 10(1):67–78

Dragani TA, Canzian F, Manenti G, Pierotti MA (1996) Hepatocarcinogenesis: a polygenic model of inherited predisposition to cancer. Tumori 82(1):1–5

El Serag HB, Mason AC (2000) Risk factors for the rising rates of primary liver cancer in the United States. Arch Intern Med 160(21):3227–3230

Feo F, De Miglio MR, Simile MM, Muroni MR, Calvisi DF, Frau M, Pascale RM (2006) Hepatocellular carcinoma as a complex polygenic disease. Interpretive analysis of recent developments on genetic predisposition. Biochim Biophys Acta 1765(2):126–147

Gaiani S, Gramantieri L, Venturoli N, Piscaglia F, Siringo S, D'Errico A, Zironi G, Grigioni W, Bolondi L (1997) What is the criterion for differentiating chronic hepatitis from compensated cirrhosis? A prospective study comparing ultrasonography and percutaneous liver biopsy. J Hepatol 27(6):979–985

Gannon C, Curley SA (2005) The role of focal liver ablation in the treatment of unresectable primary and secondary malignant liver tumors. Semin Radiat Oncol 15(4):265–272

Garcea G, Lloyd TD, Aylott C, Maddern G, Berry DP (2003) The emergent role of focal liver ablation techniques in the treatment of primary and secondary liver tumors. Eur J Cancer 39(15):2150–2164

Gruessner B (2008) Living donor organ transplantation. McGraw Hill, New York

Hinshaw L, Lee FT (2007) Cryoablation for liver. Cancer Tech Vasc Interv Rad 10(1):47–57

Ikeda K, Saitoh S, Arase Y, Chayama K, Suzuki Y, Kobayashi M, Tsubota A, Nakamura I, Murashima N, Kumada H, Kawanishi M (1999) Effect of interferon therapy on hepatocellular carcinogenesis in patients with chronic hepatitis type C: a long-term observation study of 1,643 patients using statistical bias correction with proportional hazard analysis. Hepatology 29(4):1124–1130

Kudo M, Zheng RQ, Kim SR, Okabe Y, Osaki Y, Iijima H, Itani T, Kasugai H, Kanematsu M, Ito K, Usuki N, Shimamatsu K, Kage M, Kojiro M (2008) Diagnostic accuracy of imaging for liver cirrhosis is compared to histologically proven liver cirrhosis. A multicenter collaborative study. Intervirology 51(Suppl 1):17–26

Li GZ, Speicher PJ, Lidsky ME, Darrabie MD, Scarborough JE, White RR, Turley RS, Clary BM (2014) Hepatic resection for hepatocellular carcinoma: do contemporary morbidity and mortality rates demand a transition to ablation as first-line treatment? J Am Coll Surg 218(4):827–834

Llovet JM, Real MI, Montana X, Planas R, Coll S, Aponte J, Ayuso C, Sala M, Muchart J, Sola R, Rodes J, Bruix J, Barcelona Liver Cancer Group (2002) Arterial embolisation or chemoembolisation versus symptomatic treatment in patients with unresectable hepatocellular carcinoma: a randomised controlled trial. Lancet 359(9319):1734–1739

Llovet JM, Ducreux M, Lencioni R, Di Bisceglie AM, Galle PR, Dufour JF, Greten TF, Raymond E, Roskams T, De Baere T, Ducreux M, Mazzaferro V, Bernardi M, Bruix J, Colombo M, Zhu A (2012) EASL-EORTC clinical practice guidelines: management of hepatocellular carcinoma. J Hepatol 56(4):908–943

Okanoue T, Itoh Y, Minami M, Sakamoto S, Yasui K, Sakamoto M, Nishioji K, Murakami Y, Kashima K (1999) Interferon therapy lowers the rate of progression to hepatocellular carcinoma in chronic hepatitis C but not significantly in an advanced stage: a retrospective study in 1148 patients. Viral Hepatitis Therapy Study Group. J Hepatol 30(4):653–659

Piscaglia F, Bolondi L (2007) Recent advances in the diagnosis of hepatocellular carcinoma. Hepatol Res 37(Suppl 2):S178–S192

Ribero D, Curley SA, Imamura H, Madoff DC, Nagorney DM, Ng KK, Donadon M, Vilgrain V, Torzilli G, Roh M, Vauthey JN (2008) Selection for resection of hepatocellular carcinoma and surgical strategy: indications for resection, evaluation of liver function, portal vein embolization, and resection. Ann Surgical Oncol 15(4):986–992

Sebastiani G, Ghali P, Wong P, Klein MB, Deschenes M, Myers RP (2014) Physicians' practices for diagnosing liver fibrosis in chronic liver diseases: a nationwide, Canadian survey. Can J Gastroenterol Hepatol 28(1):23–30

Skelton M, O'Neill B (2008) Targeted therapies for hepatocellular carcinoma. Clin Adv Hematol Oncol 6(3):209–218

Stevenson M, Lloyd-Jones M, Morgan MY, Wong R (2012) Non-invasive diagnostic assessment tools for the detection of liver fibrosis in patients with suspected alcohol-related liver disease: a systematic review and economic evaluation. Health Technol Assess 16(4)

Taura K, Ikai I, Hatano E, Fujii H, Uyama N, Shimahara Y (2006) Implication of frequent local ablation therapy for intrahepatic recurrence in prolonged survival of patients with hepatocellular carcinoma undergoing hepatic resection: an analysis of 610 patients over 16 years. Ann Surg 244(2):265–273

Thomas M, O'Beirne JP, Furuse J, Chan AT, Abou-Alfa G, Johnson P (2008) Systemic therapy for hepatocellular carcinoma: cytotoxic chemotherapy, targeted therapy and immunotherapy. Ann Surg Oncol 15(4):1008–1014

Trevisani F, D'Intino PE, Grazi GL, Caraceni P, Gasbarrini A, Colantoni A, Stefanini GF, Mazziotti A, Gozzetti G, Gasbarrini G, Bernardi M (1996) Clinical and pathologic features of hepatocellular carcinoma in young and older Italian patients. Cancer 77(11):2223–2232

Washburn K (2010) Model for end stage liver disease and hepatocellular carcinoma: a moving target. Transplant Rev (Orlando) 24(1):11–17

Wong TC, Lo CM (2013) Resection strategies for hepatocellular carcinoma. Semin Liver Dis 33(3):273–281

Wurmbach E, Chen YB, Khitrov G, Zhang W, Roayaie S, Schwartz M, Fiel I, Thung S, Mazzaferro V, Bruix J, Bottinger E, Friedman S, Waxman S, Llovet JM (2007) Genome-wide molecular profiles of HCV-induced dysplasia and hepatocellular carcinoma. Hepatology 45:938–947

Chapter 4
Pathophysiology of HCC

Histopathological and molecular features that lead to HCC initiation and progression are still poorly understood. There is growing evidence to suggest that gradual accumulation of mutations and genetic changes in preoneoplastic hepatocytes causes malignant transformation and leads to the development of HCC (Farazi and DePinho 2006; El-Serag and Rudolph 2007). The tumor may consist of only a single lesion or could be multiple lesions within the liver, the latter being the most common. When the tumor is well-differentiated, the neoplastic cells show features of normal hepatocytes, but in poorly differentiated tumors, the cells are large, and are often indistinguishable from those of other metastatic tumors.

The normal liver lobule is formed by hepatocytes, cholangiocytes and non-parenchymal cells (Kupffer cells, sinusoidal endothelial cells and hepatic stellate cells (HSCs)). Intrahepatic lymphocytes and liver-specific natural killer (NK) cells are present in the sinusoidal lumen and perisinusoidal space of Disse (Severi et al. 2010). Exposure to toxic substances and induction of immune responses in the liver can result in inflammation through activation of Kupffer cells and HSCs, and this can lead to necrosis (Severi et al. 2010). During this process, liver fibrosis and cirrhosis may also occur. Cirrhosis, the most advanced stage of fibrosis, is characterized by distortion of the liver parenchyma associated with septae and nodule formation, altered blood flow, and risk of liver failure (Friedman 2008). It is responsible for significant morbidity and mortality, and is one of the most important risk factors for the development of HCC (Severi et al. 2010). Although the molecular basis for the cancer-promoting effect of cirrhosis is unknown, the process of recurrent liver cell necrosis and regeneration with increased cell turnover, renders liver cells more sensitive to the adverse effects of other mutagenic agents. Both genetic and epigenetic changes may occur, eventually leading to the subsequent formation of dysplastic foci, nodules and finally hepatocellular carcinoma (Roskams and Kojiro 2010; Severi et al. 2010).

© The Author(s) 2014
R.N. Aravalli, C.J. Steer, *Hepatocellular Carcinoma*, SpringerBriefs
in Cancer Research, DOI 10.1007/978-3-319-09414-4_4

Tumor Microenvironment

Tissue environment plays a critical role in tumor formation and development (Yang et al. 2011). Carcinogenesis is a process that involves the transition of a normal cell into a preneoplastic lesion that develops into a malignant tumor (Severi et al. 2010). Interaction of different cell types in the tumor stroma with components of the extracellular matrix (ECM), such as collagen, fibronectin, laminin, glycosamino-glycans, hyaluronan, and proteoglycans, either directly or indirectly result in the acquisition of an abnormal phenotype that promotes this transformation. Tumor stroma consists of fibroblasts (also referred to as 'cancer-associated fibroblasts (CAFs)'), macrophages (liver resident Kupffer cells and other tumor-infiltrating cells), leukocytes, HSCs, endothelial cells, pericytes, neutrophils, and dendritic cells (Schrader and Iredale 2011). Each of these cells produces growth factors, cytokines, chemokines, free radicals, and other tumorigenic substrates that contribute to tumor initiation and progression (Wu et al. 2012).

Cancer-Associated Fibroblasts

Cancer-associated fibroblasts (CAFs) appear to play a critical role in tumor-stromal interactions in liver and other cancer types. In HCC, they are involved in tumor initiation and progression where they produce epidermal growth factor (EGF), hepatocyte growth factor (HGF), fibroblast growth factor (FGF), interleukin 6 (IL-6), chemokine (C-X-C motif) ligand 12 (CXCL12), and matrix metalloproteases 3 and 9 (MMP-3 and MMP-9) (Pietras and Ostman 2010). CAFs also secrete IL-8, cyclooxygenase 2 (COX-2), and secreted protein acidic rich in cysteine (SPARC) to recruit and stimulate macrophage production, which can further increase the activation of CAFs via secretion of tumor necrosis factor-α (TNF-α) and platelet-derived growth factor (PDGF) (Mueller et al. 2007; Zhang et al. 2007).

Tumor-Associated Macrophages

Tumor-associated macrophages (TAMs) are 'polarized' into M2 mononuclear phagocyte-like cells by various cytokines (IL-4, IL-10, and transforming growth factor β (TGF-β)) present in the tumor microenvironment (Mantovani et al. 2009). These M2-like TAMs, in turn, express cytokines (IL-10 and TGF-β), chemokines (CCL17, CCL22 and CCL24), vascular endothelial growth factor (VEGF), and EGF to recruit regulatory T cells (Tregs) and promote angiogenesis (Movahedi et al. 2010; Mantovani et al. 2011). Kupffer cells, resident macrophages of the liver, are liver-specific TAMs in the tumor microenvironment. These cells are able to impair $CD8^+$ cytotoxic T lymphocyte (CTL)-mediated immune responses through programmed death ligand 1(PD-L1), which interacts with programmed death 1 (PD-1), a cell surface protein of $CD8^+$ T cells (Wu et al. 2009). Moreover, when stimulated with pro-inflammatory cytokines (IL-1β, TNF-α and PDGF), Kupffer cells and HSCs produce osteopontin that plays a pivotal role in various cell signaling pathways, which promote inflammation, tumor progression and metastasis (Ramaiah and Rittling 2008; Leonardi et al. 2012).

Dendritic Cells

Dendritic cells (DCs) process antigens antigens and present them to infiltrating CTLs by expressing them on their cell surface. They possess high endocytic activity and are critical for the induction of immune surveillance in tumors, and for immune evasion (Lin et al. 2010). Such tumor-antigen specific CD8$^+$ T cell responses suppress the recurrence of HCC (Hiroishi et al. 2010). Glypican 3 (GPC3), a protein expressed by the fetal liver and downregulated in the adult liver (Iglesias et al. 2008), facilitates the binding of various growth factors to their cognizant receptors. In HCC, upregulation of GPC3 correlates with poor prognosis (Shirakawa et al. 2009). Human monocyte-derived DCs expressing a GPC3 antigen epitope were able to induce functional T cells *in vitro* and produce interferon-γ (IFN-γ) suggesting that this GPC3 epitope can be used to monitor CTL responses in patients undergoing immunotherapy (O'Beirne et al. 2010). In a recent study, infiltration of CD4$^+$/CD25$^+$ Tregs that suppress the immune responses elicited by DCs into the tumor environment of HCC patients was found to correlate with an increase in tumor size (Lee et al. 2012).

Hepatic Stellate Cells

Hepatic stellate cells are collagen-producing cells in the liver (Friedman 2008), and in response to liver injury, transdifferentiate into myofibroblast-like cells and produce cytokines, chemokines, growth factors, and ECM (Leonardi et al. 2012). This phenotypic transformation of HSCs is a key event in the development of hepatic fibrosis (Friedman 2008). HBV-encoded X protein, HCV non-structural proteins, MMP-9, PDGF, TGF-β, Janus kinase (JNK), insulin-like growth factor (IGF) binding protein 5, and cathepsins B and D are potent inducers of HSC activation and proliferation that enhance liver fibrosis and carcinogenesis (Leonardi et al. 2012).

Endothelial Cells

Endothelial cells express a variety of angiogenic receptors including VEGFR, EGFR, EGF homology domains-2 (Tie-2), PDGFR, and C-X-C chemokine receptors (CXCRs). Several signaling pathways connected with survival, proliferation, mobilization and invasion of endothelial cells are regulated by the interaction between ligands and their corresponding receptors (Leonardi et al. 2012). Tumor-associated endothelial cells express high levels of TGF-β in HCC, which act as a chemoattractant for CD105 to promote tumor angiogenesis (Benetti et al. 2008). It has been shown that CD105[+] endothelial cells express increased angiogenesis activity with greater resistance to chemotherapeutic agents and inhibitors of angiogenesis (Xiong et al. 2009).

T Cells

Infiltration of T cells into the tumor microenvironment is an important regulator of cancer progression. CD4[+]/CD25[+] Tregs outnumber CD8[+] T cells in HCC tissues compared with adjacent benign tissue. CD4[+]/CD25[+] Tregs impair cytotoxic CD8[+] T cell proliferation, activation, degranulation, and production of granzymes (A and B), and perforin (Fu et al. 2007a). In support of these findings, a number of recent studies showed that low CD8[+] T cell and high Treg numbers correlate with poor prognosis in HCC patients, especially after resection (Chen et al. 2012; Mathai et al. 2012; Wang et al. 2012). A 14-immune gene signature, which includes pro-inflammatory cytokines (TNFα- and IFN-γ) and chemokines (CXCL10, CCL5 and CCL2), has recently been identified that drives the infiltration of lymphocytes into tumor (Chew et al. 2012). This signature is an accurate predictor of patient survival at early tumor stages. Similarly, dysfunctional regulation of the immune response by excessive neutrophil activity was also reported as a poor prognostic indicator after resection of HCC (Li et al. 2011).

The Extracellular Matrix

The ECM provides structural support and anchorage to parenchymal cells and enables intracellular communication, where proteoglycans, including heparan sulfate, chondroitin sulfate, and keratan sulfates, are critical players (Leonardi et al. 2012; Wu et al. 2012). These proteoglycans facilitate the storage of various growth factors (FGF, HGF, PDGF, and VEGF) within the ECM. Of these, heparan sulfate proteoglycans are known to play an important role in the pathogenesis of HCC (Campbell et al. 2005). Collagen, the most abundant protein in the ECM, promotes cell migration and proliferation in HCC stroma (Leonardi et al. 2012). Laminins, another group of ECM proteins, are heterotrimeric components involved in various biological activities, including the assembly of basement membrane, cell attachment, migration, growth and differentiation, and angiogenesis (Miner 2008). Among these, Laminin-5 (Ln-5) is expressed in HCC nodules and its expression is associated with the metastatic phenotype of HCC (Giannelli et al. 2003). In addition, Ln-5, in conjunction with TGF-β1, promotes epithelial to mesenchymal transition (Giannelli et al. 2005). Integrins are surface receptor proteins that mediate cell-matrix and cell-cell adhesion (Leonardi et al. 2012). Overexpression of β1 integrin was found to inhibit proliferation of hepatoma cell line SMMC-7721 due to reduced cell adhesion and blockage of S-phase kinase-associated protein 2 (Skp2)-dependent degradation of p27 via the phosphoinositide 3-kinase (PI3K) pathway (Fu et al. 2007b). In contrast, overexpression of α6β4 and α3β1 integrins was found to be associated with increased migration and invasion of HCC cells in an Ln-5 dependent fashion (Bergamini et al. 2007; Kikkawa et al. 2008; Mizuno et al. 2008). Thus, each integrin appears to have a distinct role either in the promotion or prevention of HCC.

Hypoxia

Hypoxia enhances proliferation, angiogenesis, metastasis, chemo-, and radio-resistance of HCC (Wu et al. 2012). The hypoxia induced factor-1α (HIF-1α), a major transcription factor up-regulated during hypoxia, activates glycolysis via the induction of ENO1, and this activation was shown to produce an aggressive phenotype closely related to HCC metastasis (Hamaguchi et al. 2008). There is growing evidence to suggest that hypoxia is involved in neo-angiogenesis and the formation of new blood vessels, induction of gene expression through HIF-1α, and stimulation of EMT in HCC through PI3K/Akt signaling (Yan et al. 2009; Murata et al. 2010). Hypoxia also exerts profound effects on the development and evolution of the tumor microenvironment by regulating differentiation of both tumor and stromal cells (Wu et al. 2012). Moreover, during the development of liver fibrosis, HIF-1α has been shown to induce EMT in primary hepatocytes via TGF-β signaling (Copple 2010).

Under hypoxic conditions, expression of cytokines is mediated by carbonic anhydrase 9 (CA9) in Hep3B cells, suggesting that it is an important tumor marker for HCC (Abel et al. 2009). A recent study determined the gene expression pattern of HepG2 hepatoma cell line under chronic hypoxia using microarrays, and compared it with those of HCC patients. They found that a set of seven genes was differentially expressed under chronic hypoxia, including FGF21 and cyclin G2, and correlated with poor prognosis (van Malenstein et al. 2010). Hypoxia induced overexpression and intracellular accumulation of β-catenin via down-regulation of the endogenous degradation machinery in HCC cell lines (Liu et al. 2010). Bone morphogenetic proteins (BMPs) regulate embryonic hepatic development. Among these, BMP-4 is induced by hypoxia in HCC, and increased expression of BMP-4 in HCC promotes vasculogenesis and tumor progression (Maegdefrau et al. 2009). In hypoxic HCC cells, high-mobility group box 1 (HMGB1) induced expression of caspase-1 and inflammatory cytokines, which promoted tumor invasiveness and metastasis (Yan et al. 2012). Further, hypoxia-induced autophagy was shown to contribute to the chemoresistance seen in HCC cells (Song et al. 2009; Du et al. 2012).

Microenvironment plays an important role in the interplay between tumor cells and the surrounding cells such as endothelial cells (Wu et al. 2007). Any changes in its features have a major impact on tumor growth and responsiveness to therapy. Hypoxia is a common characteristic of many solid tumors, including

HCC (Toffoli and Michiels 2008). Tissue hypoxia occurs due to an imbalance between oxygen supply and consumption, due mainly to irregular blood flow in tumors. Several studies, both *in vitro* and *in vivo*, have shown that hypoxia could modulate tumor development, growth, angiogenesis, chemo-and radio-resistance (Krantz 1991; Wu et al. 2007). Understanding the regulation of hypoxia-induced erythropoietin (EPO) gene expression has provided many insights into the mechanisms of oxygen-regulated gene expression (Jelkmann 1992). EPO is a glycoprotein produced mainly in fetal liver and adult kidney and, to some extent, in adult liver. The expression of EPO is drastically altered at mRNA and protein levels under hypoxic conditions (Goldberg et al. 1987).

Early studies on the role of hypoxia in HCC were performed with cell culture models using two human hepatoma cell lines HepG2 and Hep3B (Goldberg et al. 1991). When cultured under hypoxic conditions (1 % *versus* 21 % O_2) and cobaltous chloride, both cell lines showed an increase in EPO secretion in a regulated manner (Fey and Gauldie 1990; Wenger et al. 1995). Further studies with differential screening of cDNA derived from hypoxic HepG2 showed that, in addition to hypoxia-related genes, several proinflammatory, cytokine-dependent acute phase (AP) proteins were also induced (Shweiki et al. 1992). Interleukin-6 (IL-6) has been shown to be the major mediator of the AP response in HepG2 and Hep3B cell lines *in vitro* and in rats *in vivo*, although IL-1, tumor necrosis factor α, and other cytokines are also capable of partially mediating the AP response (Shweiki et al. 1992; Goldberg and Schneider 1994). Hypoxic induction of vascular endothelial growth factor (VEGF), a potent angiogenic factor, has also been found in many different tissues and tumors (Bae et al. 1998), including hepatoma cells (Semenza 2003). In a recent study, differential display analysis was used to identify two genes that were differentially expressed under hypoxic conditions (1 % O_2, 5 % CO_2, balance N_2) in HepG2 cells. Cytochrome oxidase subunit II was downregulated in contrast to ADP/ATP translocase, and oscillin whose expression was upregulated (Tanaka et al. 2006).

Among other hypoxia-related genes, hypoxia-inducible factor 1 (HIF-1) was shown to be involved in tumor progression/metastasis and is activated in a number of cancers (Bandyopadhyay et al. 2003). Therefore, HIF-1 has been a strong candidate for targeted gene therapy. During early stages of hepatocarcinogenesis in mice and humans, HIF-1α is overexpressed in preneoplastic hepatocytic lesions. Transcriptional targets of HIF-1, such as VEGF, glut-1, c-met, and insulin-like growth factor II (IGF-II), were also overexpressed in mouse lesions. It was shown that HIF-1α was markedly down-regulated in murine HCC cell lines following treatment with a phosphatidylinositol 3-kinase (PI3K) inhibitor, LY294002, suggesting that HIF-1α expression is dependent upon PI3K/Akt signaling. Conversely, HIF-1 knockdown by siRNA resulted in decreased expression of activated Akt and that of HIF-1 target genes (Tanaka et al. 2006). NDRG1 (N-myc downstream regulated gene-1), a hypoxia-inducible protein originally described as a stress-responsive protein, is a tumor suppressor that was found to be associated with the progression of several cancers including HCC (Guan et al. 2000; Kim et al. 2002; Bandyopadhyay et al. 2004). The expression of NDRG1 was found to be markedly up-regulated in the cytoplasm and on cell membrane in human HCC. Under hypoxic stress, its cytoplasmic expression increased *in vitro*. When examined in a HCC-xenograft *in vivo*, NDRG1 expression was elevated in the cytoplasm as well as on the plasma membrane (Sibold et al. 2007).

Hypoxia is an integral part of transient and chronic HCC (Kim et al. 2002). Identification of differentially regulated genes such as those of HIF-1 signaling pathway provides us with novel targets for gene therapy. Further studies on hypoxia-induced proteins and their relationship with tumor stroma are critical for the development of novel strategies to combat hepatocarcinogenesis.

Epithelial-Mesenchymal Transition

The epithelial-mesenchymal transition (EMT) plays a central role in embryogenesis where the mesoderm generated by EMTs develops into multiple tissue types (Thiery et al. 2009). EMT is often activated during cancer progression and metastasis (Thiery et al. 2009; Iwatsuki et al. 2010), and the acquisition of EMT features has been associated with chemoresistance and recurrence of the cancer. Furthermore, the EMT can generate cells exhibiting properties of stem cells; and many types of cancer cells rely on EMT to facilitate the invasion-metastasis cascade. A number of transcription factors such as Twist, Snail, and Slug that are activated during embryogenesis, also play critical roles in tumor progression.

EMT and tumor progression of malignant *ras*-transformed hepatocytes is induced by TGF-β secreted by stromal cells in liver parenchyma and upregulation of platelet-derived growth factor (PDGF) signaling (van Zijl et al. 2009a). Increased expression and activation of Twist, Snail, VE-cadherin, and Vimentin, and downregulation of E-cadherin (CDH1) and hepatocyte nuclear factor (HNF)-4α frequently occurs and correlates with poor prognosis in liver cancer (Lee et al. 2006; Niu et al. 2007; Matsuo et al. 2009; van Zijl et al. 2009b; Yang et al. 2009). During the EMT, de-stabilization of adherent junctions occurs and cells develop properties of invasion and metastasis, as shown in mouse models of HCC (Chen et al. 2010; Ding et al. 2010; Severi et al. 2010). A recent study reported that the expression of four genes (*CDH1*, *inhibitor of DNA binding 2* (*ID2*), *MMP9*, and *transcription factor 3* (*TCF3*)) correlated with prognosis in patients with HCC and HCV (Kim et al. 2010). This four-gene signature may therefore be a basis for the molecular classification of HCC.

Oxidative Stress

In general, all life forms preserve a reducing cellular redox environment within their cells by maintaining a constant input of metabolic energy, a process controlled by enzymes with varying cellular functions. Any imbalance in the normal redox state can have toxic effects in the cell due to an increased production of free radicals (ROS and reactive nitrogen species (RNS)) and peroxides, resulting in oxidative stress (Severi et al. 2010). In the liver, mitochondrial and cytochrome P450 enzymes of hepatocytes, endotoxin-activated Kupffer cells, and neutrophils represent the main sources of free radicals (Okuda et al. 2002). ROS and RNS are involved in the transcription and activation of cytokines and growth factors, thereby playing crucial roles in the pathogenesis and the progression of HCC (Marra et al. 2011). For instance, increased expression of 8-hydroxy-2′-deoxy-guanosine (8-OHdG) correlated with a high risk of developing HCC underscoring the notion that oxidative DNA damage by ROS leads to hepatocarcinogenesis (Chuma et al. 2008; Tanaka et al. 2008).

Cells have developed mechanisms to counteract the detrimental effects of oxidative stress that include redox active glutathione (GSH), and thioredoxin (TRX), and anti-oxidant enzymes superoxide dismutase (SOD), catalase and glutathione peroxidase (GPx) (Severi et al. 2010). GSH is involved in cell proliferation, and changes in cellular levels of GSH/glutathione disulfide (GSSG) were shown to cause differentiation and apoptosis due to the induction of oxidative stress (Nkabyo et al. 2002). Malondialdehyde (MDA) formed during lipid peroxidation was shown to accumulate in the serum of chronic hepatitis patients, serving as a potential biomarker for HCC (Nagoev et al. 2002). Due to the cellular imbalance caused by ROS production, highly reactive radicals can damage DNA, RNA, proteins and lipid components, and this may lead to mutations or apoptosis (Shimoda et al. 1994). In HBV-associated HCC patients, levels of superoxide radicals, MDA and GSSG, and the activities of redox enzymes superoxide dismutase (SOD) and glutathione reductase (GRx) were found to be significantly higher than in healthy individuals (Abel et al. 2009; Tsai et al. 2009). In these patients, disruption of the redox balance could create an environment for resistance to oxidative stress in tumor tissues; and alterations in membrane cholesterol, phospholipids, fatty acids, and low lipid peroxidation are likely to be important determinants underlying the selective growth advantage of HCC cells (Abel et al. 2009).

Cancer Stem Cells

During early tumor development, a single or few cells transform and begin to grow continuously as subpopulations of mutated cells which ultimately drive tumor growth and progression (Heppner and Miller 1998; Merlo et al. 2006). In this "stochastic or clonal evolution model", single mutated cells can possess unlimited proliferative potential to form tumors and develop resistance to therapy. This paradigm has been challenged by the hypothesis that a small population of quiescent cells with 'stem-like characteristics' drive tumor growth, recurrence, and resistance to chemo- and radiotherapy (Reya et al. 2001). In this "hierarchical or cancer stem cell (CSC) model", only a few cells possess the potential for self-renewal, generation of heterogeneity within the tumor due to their multipotent nature, and the ability to recapitulate the original tumor. The tumor-initiating ability of CSCs was first demonstrated in a mouse human breast cancer model (Al-Hajj et al. 2003). Subsequently, CSCs have been identified and were successfully isolated from a number of solid tumors, including HCC (Suetsugu et al. 2006; Ma et al. 2007; Visvader and Lindeman 2008).

The origin of HCC from human progenitor cells (HPCs) was inferred from the observation that most HCC specimens contain a mixture of mature cells and those phenotypically similar to HPCs and expressing ovalbumin 6 (OV-6), cytokeratin 7 (CK7) and 19, and chromogranin-A (Libbrecht and Roskams 2002). Later, cells resembling HPCs were described in hepatoblastoma and were shown to contain both HCC component and CCA characteristics, suggesting origin from a bipotential progenitor cell (Theise et al. 2003). Initial evidence that CSCs might contribute to the development of HCC came from a side-population (SP) of cells with stem-like characteristics from the HCC cell lines HuH7, and PLC5 (Chiba et al. 2006). These cells were highly proliferative and formed tumors in NOD/SCID mice, but the non-SP population was not tumorigenic, suggesting that CSCs may have a critical role in the formation of HCC. HuH7-derived SP cells in the G0 phase of cell cycle were highly tumorigenic in NOD/SCID mice with self-renewal and bi-lineage differentiation potential (Kamohara et al. 2008). SP cells with stem-like features have also been isolated from several other HCC cell lines including, HCCLM3, MHCC97-H, MHCC97-L and Hep3B (Shi et al. 2008).

CD133$^+$ cells from HuH7 and PLC5 cultures as well as from primary tumor samples from SCID mice possessed higher tumorigenicity and clonogenicity than CD133$^-$ cells (Suetsugu et al. 2006; Yin et al. 2007). These CD133$^+$ CSCs were resistant to chemotherapeutic agents (doxorubicin and 5-fluorouracil) due to the upregulation of ATP-binding cassette (ABC) superfamily transporters (Zhu et al. 2010) through activation of Akt/PKB and Bcl-2 cell survival pathways (Ma et al. 2008a). They were also shown to be resistant to radiotherapy due to their reduced expression of reactive oxygen species (ROS) production and increased activation of mitogen-activated protein kinase (MAP)/PI3K signaling (Piao et al. 2012). In general, CD133$^+$ cells isolated from human HCC cell lines represented only a minority of the tumor cell population, but they possessed a greater colony-forming efficiency, and higher proliferation potential and ability to form tumors in animal models (Ma et al. 2007).

Efforts to further characterize CD133$^+$ cells have led to the identification of CD44 and alcohol dehydrogenase 1A1 (ALDH) as putative markers for CSCs. Cells expressing CD133 and CD90 were resistant to doxorubicin and carboplatin *in vitro* (Tomuleasa et al. 2010), and those expressing CD133 and CD44 possessed a high tumorigenic capacity when xenografted in mice than CD133$^+$/CD44$^-$ cells (Zhu et al. 2010). Expression of both CD133 and ALDH was associated with greater tumorigenicity than their CD133$^-$/ALDH$^+$ or CD133$^-$/ALDH$^-$ counterparts, both *in vitro* and *in vivo* (Ma et al. 2008b). More recently, CD44$^+$ CSCs have been identified in canine HCC and cholangiocarcinoma samples (Cogliati et al. 2010).

Several studies suggest that HCC with stem cell features have a poor prognosis. While tumors expressing cytokeratins 10 and 19 (CK10 and CK19) were associated with HCC invasiveness and displayed poor prognosis after resection

(Uenishi et al. 2003; Yang et al. 2008), other markers such as α-fetoprotein (AFP) and epithelial cell adhesion molecule (EpCAM) were also found to be of prognostic value. The combination of EpCAM$^+$ and AFP$^+$ predicted poor survival, in contrast to EpCAM$^-$/AFP$^-$ which was associated with good prognosis (Yamashita et al. 2008, 2009). In another study, Yang et al. described the presence of CD45$^-$/CD90$^+$ cells in liver tumors and the peripheral blood of patients with HCC but not in normal or cirrhotic patients. Serial transplantation of these CD90$^+$ cells resulted in the formation of tumor nodules in immunodeficient mice (Yang et al. 2008). These investigators further showed that CD45$^-$/CD90$^+$ cells, but not CD90$^-$ cells, from HCC cell lines had tumorigenic potential (Yang et al. 2008). The gene expression profile of these CD45$^-$/CD90$^+$ cells indicated that they also had stem cell-like phenotype. However, no correlation was found between expression of CD133$^+$, CD133$^+$/CD44$^+$ and CD133$^+$/CD24$^-$ and the patients' clinical-pathological status or with the cancer stem cell marker-positive cells in tissue specimens of liver metastases (Salnikov et al. 2009).

Inflammation

Liver inflammation is known to occur with exposure to a variety of agents including, viruses, bacteria, metabolites of alcohol, drugs, and chemicals (Severi et al. 2010; Fallot et al. 2012). If hepatic metabolism is impaired to some degree and fails to convert drugs and chemicals to non-reactive or non-immunogenic substances, the metabolic intermediates formed in hepatic tissues may cause liver damage. In such cases, Kupffer and other cell types release cytokines and chemokines that result in inflammation of the liver. This reaction coupled with deregulated hepatocyte proliferation can contribute to the pathogenesis of liver cancer (Maeda et al. 2005; Elsharkawy and Mann 2007).

The liver contains various cell types that produce cytokines and chemokines as well as those susceptible to the actions of these immune mediators. Hepatocytes express cell surface receptors for a number of cytokines such as IL-1β, IL-6, and TNF-α. Liver sinusoidal endothelial cells are both targets and a source of various cytokines (Leonardi et al. 2012), many of which are also produced by Kupffer cells. These cells also express and release an important pro-inflammatory cytokine IL-6 which is associated with increased risk for HCC, particularly in the presence of cirrhosis (Nakagawa et al. 2009; Wong et al. 2009). It has been well documented that IL-6 is also involved in tumor cell proliferation by inhibiting apoptosis through activation of signal transducer and activator of transcription 3 (STAT3) (Yu et al. 2009). IL-6 also forms a critical link between obesity and cancer, and the gender disparity seen during HCC development is attributed to the suppression of IL-6 production in Kupffer cells by estrogen (Naugler et al. 2007; Park et al. 2007).

Another pro-inflammatory immune mediator, TNF-α, is produced by Kupffer cells and other immune cells in response to tissue injury, modulates NF-κB and Akt pathways, and is involved in tumor development and progression (Leonardi et al. 2012). It also induces oxidative stress by causing DNA damage through the formation of 8-oxo-deoxyguanosine (8-oxodG) in primary murine hepatocytes (Wheelhouse et al. 2003). Interestingly, the role of TNF-α expression in HCC remains controversial with varying reports of expression from high (Nakazaki 1992) to low (Bortolami et al. 2002; Leonardi et al. 2012).

IL-1β, a key pro-inflammatory cytokine in HCC, promotes HSC proliferation, activation, and transdifferentiation into the myofibroblastic phenotype. It also stimulates HSCs to produce and activate MMPs, in particular MMP-9 (Han et al. 2004). IL-1β has also been shown to induce the expression of TNF-related apoptosis-inducing ligand (TRAIL) in HCC cell lines (HepG2, Hep3B, HuH7) (Leonardi et al. 2012).

In patients with HBV-positive metastatic HCC, a global Th1/Th2-like cytokine shift takes place in which a Th2-like cytokine profile (IL-4, IL-8, IL-10, and IL-5) is significantly elevated, with the reduced expression of Th1-like cytokines (IL-1α, IL-1β, IL-2, IL-12p35, IL-12p40, IL-15, TNF-α, and IFN-γ) (Budhu et al. 2006). This Th1/Th2 cytokine profile was found to be associated with the metastatic phenotype, and suggested that a shift toward anti-inflammatory/immune-suppressive responses may promote HCC metastases.

IL-12 acts as a tumor suppressor by inducing the production of interferon-γ (IFN-γ) in natural killer (NK) cells and naïve T cells. It also promotes helper T cell differentiation, enhancing cell-mediated immune responses, and by activating tumor-specific CTLs (Trinchieri 2003). High levels of IL-12 expression have been detected in the serum of patients with chronic hepatitis, liver cirrhosis, and HCC (Kitaoka et al. 2009). When IL-12 was expressed from an adenoviral vector in rat liver treated with DEN, tumor growth was inhibited in 60 % of animals due to the activation of NK cells and inhibition of angiogenesis suggesting that IL-12 may be an effective anti-tumor therapy for HCC (Barajas et al. 2001). Similar results were reported in a mouse model of HCC (Yamashita et al. 2001), where intratumoral injection of IL-12 induced lymphocyte infiltration by NK cells, CD3[+] cells, and Mac-1 positive cells into the tumor and reduced angiogenesis thereby inhibiting tumor growth. However, the clinical use of IL-12 is limited due to the severe systemic toxicity from elevated IFN-γ levels with high dose therapy versus minimal efficacy with low dose (Leonardi et al. 2012).

Several reports have been published on the single nucleotide polymorphisms (SNPs) of various cytokines. It has been demonstrated that the C31T polymorphism in IL-1β may be a genetic marker for the development of hepatitis-related HCC (Wang et al. 2003). In addition, C626G and A683C polymorphisms of TRAIL receptor I were found to be associated with increased risk of developing HCV-associated HCC (Körner et al. 2012). Interestingly, IL-28B SNPs are associated with spontaneous and treatment-induced elimination of HCV. However, these IL28B SNPs appear to provide protection against inflammation and fibrosis in HCV patients, due to poor response to interferon, and were not predictive of HCC development (Marabita et al. 2011; Bochud et al. 2012; Joshita et al. 2012).

References

Abel S, De Kock M, van Schalkwyk DJ, Swanevelder S, Kew MC, Gelderblom WC (2009) Altered lipid profile, oxidative status and hepatitis B virus interactions in human hepatocellular carcinoma. Prostaglandins Leukot Essent Fatty Acids 81(5–6):391–399

Al-Hajj M, Wicha MS, Benito-Hernandez A, Morrison SJ, Clarke MF (2003) Prospective identification of tumorigenic breast cancer cells. Proc Natl Acad Sci U S A 100(7):3983–3988

Bae MK, Kwon YW, Kim MS, Bae SK, Bae MH, Lee YM, Kim YJ, Kim KW (1998) Identification of genes differentially expressed by hypoxia in hepatocellular carcinoma cells. Biochem Biophys Res Commun 243(1):158–162

Bandyopadhyay S, Pai SK, Gross SC, Hirota S, Hosobe S, Miura K, Saito K, Commes T, Hayashi S, Watabe M, Watabe K (2003) The Drg-1 gene suppresses tumor metastasis in prostate cancer. Cancer Res 63(8):1731–1736

Bandyopadhyay S, Pai SK, Hirota S, Hosobe S, Takano Y, Saito K, Piquemal D, Commes T, Watabe M, Gross SC, Wang Y, Ran S, Watabe K (2004) Role of the putative tumor metastasis suppressor gene Drg-1 in breast cancer progression. Oncogene 22(33):5675–5681

Barajas M, Mazzolini G, Genové G, Bilbao R, Narvaiza I, Schmitz V, Sangro B, Melero I, Qian C, Prieto J (2001) Gene therapy of orthotopic hepatocellular carcinoma in rats using adenovirus coding for interleukin 12. Hepatology 33(1):52–61

Benetti A, Berenzi A, Gambarotti M, Garrafa E, Gelati M, Dessy E, Portolani N, Piardi T, Giulini SM, Caruso A, Invernici G, Parati EA, Nicosia R, Alessandri G (2008) Transforming growth factor-β1 and CD105 promote the migration of hepatocellular carcinoma–derived endothelium. Cancer Res 68(20):8626–8634

Bergamini C, Sgarra C, Trerotoli P, Lupo L, Azzariti A, Antonaci S, Giannelli G (2007) Laminin-5 stimulates hepatocellular carcinoma growth through a different function of α6β4 and α3β1 integrins. Hepatology 46(6):1801–1809

Bochud PY, Bibert S, Kutalik Z, Patin E, Guergnon J, Nalpas B, Goossens N, Kuske L, Müllhaupt B, Gerlach T, Heim MH, Moradpour D, Cerny A, Malinverni R, Regenass S, Dollenmaier G, Hirsch H, Martinetti G, Gorgiewski M, Bourlière M, Poynard T, Theodorou I, Abel L, Pol S, Dufour JF, Negro F (2012) IL28B alleles associated with poor hepatitis C virus (HCV) clearance protect against inflammation and fibrosis in patients infected with non-1 HCV genotypes. Hepatology 55(2):384–394

Bortolami M, Venturi C, Giacomelli L, Scalerta R, Bacchetti S, Marino F, Floreani A, Lise M, Naccarato R, Farinati F (2002) Cytokine, infiltrating macrophage and T cell-mediated response to development of primary and secondary human liver cancer. Dig Liver Dis 34(11):794–801

Budhu A, Forgues M, Ye QH, Jia HL, He P, Zanetti KA, Kammula US, Chen Y, Qin LX, Tang ZY, Wang XW (2006) Prediction of venous metastases, recurrence and prognosis in hepatocellular carcinoma based on a unique immune response signature of the liver microenvironment. Cancer Cell 10(2):99–111

Campbell JS, Hughes SD, Gilbertson DG, Palmer TE, Holdren MS, Haran AC, Odell MM, Bauer RL, Ren HP, Haugen HS, Yeh MM, Fausto N (2005) Platelet-derived growth factor C induces liver fibrosis, steatosis, and hepatocellular carcinoma. Proc Natl Acad Sci U S A 102(9): 3389–3394

Chen L, Chan TH, Yuan YF, Hu L, Huang J, Ma S, Wang J, Dong SS, Tang KH, Xie D, Li Y, Guan XY (2010) CHD1L promotes hepatocellular carcinoma progression and metastasis in mice and is associated with these processes in human patients. J Clin Invest 120(4):1178–1191

Chen KJ, Zhou L, Xie HY, Ahmed TE, Feng XW, Zheng SS (2012) Intratumoral regulatory T cells alone or in combination with cytotoxic T cells predict prognosis of hepatocellular carcinoma after resection. Med Oncol 29(3):1817–1826

Chew V, Chen J, Lee D, Loh E, Lee J, Lim KH, Weber A, Slankamenac K, Poon RT, Yang H, Ooi LL, Toh HC, Heikenwalder M, Ng IO, Nardin A, Abastado JP (2012) Chemokine-driven lymphocyte infiltration: an early intratumoural event determining long-term survival in resectable hepatocellular carcinoma. Gut 61(3):427–438

Chiba T, Kita K, Zheng YW, Yokosuka O, Saisho H, Iwama A, Nakauchi H, Taniguchi H (2006) Side population purified from hepatocellular carcinoma cells harbors cancer stem cell-like properties. Hepatology 44(1):240–251

Chuma M, Hige S, Nakanishi M, Ogawa K, Natsuizaka M, Yamamoto Y, Asaka M (2008) 8-Hydroxy-2'-deoxy-guanosine is a risk factor for development of hepatocellular carcinoma in patients with chronic hepatitis C virus infection. J Gastroenterol Hepatol 23(9): 1431–1436

Cogliati B, Aloia TP, Bosch RV, Alves VA, Hernandez-Blazquez FJ, Dagli ML (2010) Identification of hepatic stem/progenitor cells in canine hepatocellular and cholangiocellular carcinoma. Vet Comp Oncol 8(2):112–121

Copple BL (2010) Hypoxia stimulates hepatocyte epithelial to mesenchymal transition by hypoxia inducible factor- and transforming growth factor-β-dependent mechanisms. Liver Int 30(5):669–682

Ding W, You H, Dang H, LeBlanc F, Galicia V, Lu SC, Stiles B, Rountree CB (2010) Epithelial-to-mesenchymal transition of murine liver tumor cells promotes invasion. Hepatology 52(3): 945–953

Du H, Yang W, Chen L, Shen B, Peng C, Li H, Ann DK, Yen Y, Qiu W (2012) Emerging role of autophagy during ischemia-hypoxia and reperfusion in hepatocellular carcinoma. Int J Oncol 40(6):2049–2057

El-Serag HB, Rudolph KL (2007) Hepatocellular carcinoma: epidemiology and molecular carcinogenesis. Gastroenterology 132(7):2557–2576

Elsharkawy AM, Mann DA (2007) Nuclear factor-kB and the hepatic inflammation-fibrosis-cancer axis. Hepatology 46(2):590–597

Fallot G, Neuveut C, Buendia MA (2012) Diverse roles of hepatitis B virus in liver cancer. Curr Opin Virol 2(4):467–473

Farazi PA, DePinho RA (2006) Hepatocellular carcinoma pathogenesis: from genes to environment. Nat Rev Cancer 6(9):674–687

Fey GH, Gauldie J (1990) The acute phase response of the liver in inflammation. Prog Liver Dis 9:89–116

Friedman SL (2008) Mechanisms of hepatic fibrogenesis. Gastroenterology 134(6):1655–1669

Fu J, Xu D, Liu Z, Shi M, Zhao P, Fu B, Zhang Z, Yang H, Zhang H, Zhou C, Yao J, Jin L, Wang H, Yang Y, Fu YX, Wang FS (2007a) Increased regulatory T cells correlate with CD8 T-cell impairment and poor survival in hepatocellular carcinoma patients. Gastroenterology 132(7):2328–2339

Fu Y, Fang Z, Liang Y, Zhu X, Prins P, Li Z, Wang L, Sun L, Jin J, Yang Y, Zha X (2007b) Overexpression of integrin beta1 inhibits proliferation of hepatocellular carcinoma cell SMMC-7721 through preventing Skp2-dependent degradation of p27 via PI3K pathway. J Cell Biochem 102(3):704–718

Giannelli G, Fransvea E, Bergamini C, Marinosci F, Antonaci S (2003) Laminin-5 chains are expressed differentially in metastatic and nonmetastatic hepatocellular carcinoma. Clin Cancer Res 9(10 Pt 1):3684–3691

Giannelli G, Bergamini C, Fransvea E, Sgarra C, Antonaci S (2005) Laminin-5 with transforming growth factor- β1 induces epithelial to mesenchymal transition in hepatocellular carcinoma. Gastroenterology 129(5):1375–1383

Goldberg MA, Schneider TJ (1994) Similarities between the oxygen-sensing mechanisms regulating the expression of vascular endothelial growth factor and erythropoietin. J Biol Chem 269:4355–4359

Goldberg MA, Glass GA, Cunnigham JM, Bunn HF (1987) The regulated expression of erythropoietin by two human hepatoma cell lines. Proc Natl Acad Sci USA 84:7972–7976

Goldberg MA, Gaut CC, Bunn HF (1991) Erythropoietin mRNA levels are governed by both the rate of gene transcription and posttranscriptional events. Blood 77:271–277

Guan RJ, Ford HL, Fu Y, Li Y, Shaw LM, Pardee AB (2000) Drg-1 as a differentiation-related, putative metastatic suppressor gene in human colon cancer. Cancer Res 60:749–755

Hamaguchi T, Iizuka N, Tsunedomi R, Hamamoto Y, Miyamoto T, Iida M, Tokuhisa Y, Sakamoto K, Takashima M, Tamesa T, Oka M (2008) Glycolysis module activated by hypoxia-inducible factor 1α is related to the aggressive phenotype of hepatocellular carcinoma. Int J Oncol 33(4):725–731

Han YP, Zhou L, Wang J, Xiong S, Garner WL, French SW, Tsukamoto H (2004) Essential role of matrix metalloproteinases in interleukin-1-induced myofibroblastic activation of hepatic stellate cell in collagen. J Biol Chem 279(6):4820–4828

Heppner GH, Miller FR (1998) The cellular basis of tumor progression. Int Rev Cytol 177:1–56

Hiroishi K, Eguchi J, Baba T, Shimazaki T, Ishii S, Hiraide A, Sakaki M, Doi H, Uozumi S, Omori R, Matsumura T, Yanagawa T, Ito T, Imawari M (2010) Strong CD8$^+$ T-cell responses against tumor-associated antigens prolong the recurrence-free interval after tumor treatment in patients with hepatocellular carcinoma. J Gastroenterol 45(4):451–458

Iglesias BV, Centeno G, Pascuccelli H, Ward F, Peters MG, Filmus J, Puricelli L, de Kier Joffé EB (2008) Expression pattern of glypican-3 (GPC3) during human embryonic and fetal development. Histol Histopathol 23(11):1333–1340

Iwatsuki M, Mimori K, Yokobori T, Ishi H, Beppu T, Nakamori S, Baba H, Mori M (2010) Epithelial–mesenchymal transition in cancer development and its clinical significance. Cancer Sci 101(2):293–299

Jelkmann W (1992) Erythropoietin: structure, control of production, and function. Physiol Rev 72:449–489

Joshita S, Umemura T, Katsuyama Y, Ichikawa Y, Kimura T, Morita S, Kamijo A, Komatsu M, Ichijo T, Matsumoto A, Yoshizawa K, Kamijo N, Ota M, Tanaka E (2012) Association of IL28B gene polymorphism with development of hepatocellular carcinoma in Japanese patients with chronic hepatitis C virus infection. Hum Immunol 73(3):298–300

Kamohara Y, Haraguchi N, Mimori K, Tanaka F, Inoue H, Mori M, Kanematsu T (2008) The search for cancer stem cells in hepatocellular carcinoma. Surgery 144(2):119–124

Kikkawa Y, Sudo R, Kon J, Mizuguchi T, Nomizu M, Hirata K, Mitaka T (2008) Laminin α5 mediates ectopic adhesion of hepatocellular carcinoma through integrins and/or Lutheran/basal cell adhesion molecule. Exp Cell Res 314(14):2579–2590

Kim KR, Moon HE, Kim KW (2002) Hypoxia-induced angiogenesis in human hepatocellular carcinoma. J Mol Med 80:703–714

Kim J, Hong SJ, Park JY, Park JH, Yu YS, Park SY, Lim EK, Choi KY, Lee EK, Paik SS, Lee KG, Wang HJ, Do IG, Joh JW, Kim DS (2010) Epithelial-mesenchymal transition gene signature to predict clinical outcome of hepatocellular carcinoma. Cancer Sci 101(6):1521–1528

Kitaoka S, Shiota G, Kawasaki H (2009) Serum levels of interleukin-10, interleukin-12 and soluble interleukin-2 receptor in chronic liver disease type C. Hepatogastroenterology 50(53):1569–1574

Körner C, Riesner K, Krämer B, Eisenhardt M, Glässner A, Wolter F, Berg T, Müller T, Sauerbruch T, Nattermann J, Spengler U, Nischalke HD (2012) TRAIL receptor I (DR4) polymorphisms C626G and A683C are associated with an increased risk for hepatocellular carcinoma (HCC) in HCV-infected patients. BMC Cancer 12:85

Krantz SB (1991) Erythropoietin. Blood 77:419–434

Lee TK, Poon RT, Yuen AP, Ling MT, Kwok WK, Wang XH, Wong YC, Guan XY, Man K, Chau KL, Fan ST (2006) Twist overexpression correlates with hepatocellular carcinoma metastasis through induction of epithelial-mesenchymal transition. Clin Cancer Res 12(18):5369–5376

Lee WC, Wu TJ, Chou HS, Yu MC, Hsu PY, Hsu HY, Wang CC (2012) The impact of CD4$^+$CD25$^+$ T cells in the tumor microenvironment of hepatocellular carcinoma. Surgery 151(2):213–222

Leonardi GC, Candido S, Cervello M, Nicolosi D, Raiti F, Travali S, Spandidos DA, Libra M (2012) The tumor microenvironment in hepatocellular carcinoma (review). Int J Oncol 40(6):1733–1747

Li YW, Qiu SJ, Fan J, Zhou J, Gao Q, Xiao YS, Xu YF (2011) Intratumoral neutrophils: a poor prognostic factor for hepatocellular carcinoma following resection. J Hepatol 54(3):497–505

Libbrecht L, Roskams T (2002) Hepatic progenitor cells in human liver diseases. Semin Cell Dev Biol 13(6):389–396

Lin A, Schildknecht A, Nguyen LT, Ohashi PS (2010) Dendritic cells integrate signals from the tumor microenvironment to modulate immunity and tumor growth. Immunol Lett 127(2):77–84

Liu L, Zhu XD, Wang WQ, Shen Y, Qin Y, Ren ZG, Sun HC, Tang ZY (2010) Activation of β-catenin by hypoxia in hepatocellular carcinoma contributes to enhanced metastatic potential and poor prognosis. Clin Cancer Res 16(10):2740–2750

Ma S, Chan KW, Hu L, Lee TK, Wo JY, Ng IO, Zheng BJ, Guan XY (2007) Identification and characterization of tumorigenic liver cancer stem/progenitor cells. Gastroenterology 132(7):2542–2556

Ma S, Lee TK, Zheng BJ, Chan KW, Guan XY (2008a) CD133$^+$ HCC cancer stem cells confer chemoresistance by preferential expression of the Akt/PKB survival pathway. Oncogene 27(12):1749–1758

Ma S, Chan KW, Lee TK, Tang KH, Wo JY, Zheng BJ, Guan XY (2008b) Aldehyde dehydrogenase discriminates the CD133 liver cancer stem cell populations. Mol Cancer Res 6(7):1146–1153

Maeda S, Kamata H, Luo JL, Leffert H, Karin M (2005) IKKβ couples hepatocyte death to cytokine-driven compensatory proliferation that promotes chemical hepatocarcinogenesis. Cell 121(7):977–990

Maegdefrau U, Amann T, Winklmeier A, Braig S, Schubert T, Weiss TS, Schardt K, Warnecke C, Hellerbrand C, Bosserhoff AK (2009) Bone morphogenetic protein 4 is induced in hepatocellular carcinoma by hypoxia and promotes tumour progression. J Pathol 218(4): 520–529

Mantovani A, Sica A, Allavena P, Garlanda C, Locati M (2009) Tumor-associated macrophages and the related myeloid-derived suppressor cells as a paradigm of the diversity of macrophage activation. Hum Immunol 70(5):325–330

Mantovani A, Germano G, Marchesi F, Locatelli M, Biswas SK (2011) Cancer-promoting tumor-associated macrophages: new vistas and open questions. Eur J Immunol 41(9):2522–2525

Marabita F, Aghemo A, De Nicola S, Rumi MG, Cheroni C, Scavelli R, Crimi M, Soffredini R, Abrignani S, De Francesco R, Colombo M (2011) Genetic variation in the interleukin-28B gene is not associated with fibrosis progression in patients with chronic hepatitis C and known date of infection. Hepatology 54(4):1127–1134

Marra M, Sordelli IM, Lombardi A, Lamberti M, Tarantino L, Giudice A, Stiuso P, Abbruzzese A, Sperlongano R, Accardo M, Agresti M, Caraglia M, Sperlongano P (2011) Molecular targets and oxidative stress biomarkers in hepatocellular carcinoma: an overview. J Transl Med 9:171

Mathai AM, Kapadia MJ, Alexander J, Kernochan LE, Swanson PE, Yeh MM (2012) Role of Foxp3-positive tumor-infiltrating lymphocytes in the histologic features and clinical outcomes of hepatocellular carcinoma. Am J Surg Pathol 36(7):980–986

Matsuo N, Shiraha H, Fujikawa T, Takaoka N, Ueda N, Tanaka S, Nishina S, Nakanishi Y, Uemura M, Takaki A, Nakamura S, Kobayashi Y, Nouso K, Yagi T, Yamamoto K (2009) Twist expression promotes migration and invasion in hepatocellular carcinoma. BMC Cancer 9:240

Merlo LM, Pepper JW, Reid BJ, Maley CC (2006) Cancer as an evolutionary and ecological process. Nat Rev Cancer 6(12):924–935

Miner JH (2008) Laminins and their roles in mammals. Microsc Res Tech 71(5):349–356

Mizuno H, Ogura M, Saito Y, Sekine W, Sano R, Gotou T, Oku T, Itoh S, Katabami K, Tsuji T (2008) Changes in adhesive and migratory characteristics of hepatocellular carcinoma (HCC) cells induced by expression of α3β1 integrin. Biochim Biophys Acta 1780:564–570

Movahedi K, Laoui D, Gysemans C, Baeten M, Stangé G, Van den Bossche J, Mack M, Pipeleers D, In't Veld P, De Baetselier P, Van Ginderachter JA (2010) Different tumor microenvironments contain functionally distinct subsets of macrophages derived from Ly6C(high) monocytes. Cancer Res 71(14):5728–5739

Mueller L, Goumas FA, Affeldt M, Sandtner S, Gehling UM, Brilloff S, Walter J, Karnatz N, Lamszus K, Rogiers X, Broering DC (2007) Stromal fibroblasts in colorectal liver metastases originate from resident fibroblasts and generate an inflammatory microenvironment. Am J Pathol 171(5):1608–1618

Murata K, Suzuki H, Okano H, Oyamada T, Yasuda Y, Sakamoto A (2010) Hypoxia-induced des-γ-carboxy prothrombin production in hepatocellular carcinoma. Int J Oncol 36(1):161–170

Nagoev BS, Abidov MT, Ivanova MR (2002) LPO and free-radical oxidation parameters in patients with acute viral hepatitis. Bull Exp Biol Med 134(6):557–558

Nakagawa H, Maeda S, Yoshida H, Tateishi R, Masuzaki R, Ohki T, Hayakawa Y, Kinoshita H, Yamakado M, Kato N, Shiina S, Omata M (2009) Serum IL-6 levels and the risk for hepatocarcinogenesis in chronic hepatitis C patients: an analysis based on gender differences. Int J Cancer 125(10):2264–2269

Nakazaki H (1992) Preoperative and postoperative cytokines in patients with cancer. Cancer 70(3):709–713

Naugler WE, Sakurai T, Kim S, Maeda S, Kim K, Elsharkawy AM, Karin M (2007) Gender disparity in liver cancer due to sex differences in MyD88-dependent IL-6 production. Science 317(5834):121–124

Nkabyo YS, Ziegler TR, Gu LH, Watson WH, Jones DP (2002) Glutathione and thioredoxin redox during differentiation in human colon epithelial (Caco-2) cells. Am J Physiol Gastrointest Liver Physiol 283(6):G1352–G1359

Niu RF, Zhang L, Xi GM, Wei XY, Yang Y, Shi YR, Hao XS (2007) Up-regulation of twist induces angiogenesis and correlates with metastasis in hepatocellular carcinoma. J Exp Clin Cancer Res 26(3):385–394

O'Beirne J, Farzaneh F, Harrison PM (2010) Generation of functional CD8[+] T cells by human dendritic cells expressing glypican-3 epitopes. J Exp Clin Cancer Res 29(1):48

Okuda M, Li K, Beard MR, Showalter LA, Scholle F, Lemon SM, Weinman SA (2002) Mitochondrial injury, oxidative stress, and antioxidant gene expression are induced by hepatitis C virus core protein. Gastroenterology 122(2):366–375

Park EJ, Lee JH, Yu GY, He G, Ali SR, Holzer RG, Osterreicher CH, Takahashi H, Karin M (2007) Dietary and genetic obesity promote liver inflammation and tumorigenesis by enhancing IL-6 and TNF expression. Cell 140(2):197–208

Piao LS, Hur W, Kim TK, Hong SW, Kim SW, Choi JE, Sung PS, Song MJ, Lee BC, Hwang D, Yoon SK (2012) CD133[+] liver cancer stem cells modulate radioresistance in human hepatocellular carcinoma. Cancer Lett 315(2):129–137

Pietras K, Ostman A (2010) Hallmarks of cancer: interactions with the tumor stroma. Exp Cell Res 316(8):1324–1331

Ramaiah SK, Rittling S (2008) Pathophysiological role of osteopontin in hepatic inflammation, toxicity, and cancer. Toxicol Sci 103(1):4–13

Reya T, Morrison SJ, Clarke MF, Weissman IL (2001) Stem cells, cancer, and cancer stem cells. Nature 414(6859):105–111

Roskams T, Kojiro M (2010) Pathology of early hepatocellular carcinoma: conventional and molecular diagnosis. Semin Liver Dis 31(1):17–25

Salnikov AV, Groth A, Apel A, Kallifatidis G, Beckermann BM, Khamidjanov A, Ryschich E, Buchler MW, Herr I, Moldenhaur G (2009) Targeting of cancer stem cell marker EpCAM by bispecific antibody EpCAMxCD3 inhibits pancreatic carcinoma. J Cell Mol Med 13(9B): 4023–4033

Schrader J, Iredale JP (2011) The inflammatory microenvironment of HCC – the plot becomes complex. J Hepatol 54(5):853–855

Semenza GL (2003) Targeting HIF-1 for cancer therapy. Nat Rev Cancer 3:721–732

Severi T, van Malenstein H, Verslype C, van Pelt JF (2010) Tumor initiation and progression in hepatocellular carcinoma: risk factors, classification, and therapeutic targets. Acta Pharmacol Sin 31(11):1409–1420

Shi GM, Xu Y, Fan J, Zhou J, Yang XR, Qiu SJ, Liao Y, Wu WZ, Ji Y, Ke AW, Ding ZB, He YZ, Wu B, Yang GH, Qin WZ, Zhang W, Zhu J, Min ZH, Wu ZQ (2008) Identification of side population cells in human hepatocellular carcinoma cell lines with stepwise metastatic potentials. J Cancer Res Clin Oncol 134(11):1155–1163

Shimoda R, Nagashima M, Sakamoto M, Yamaguchi N, Hirohashi S, Yokota J, Kasai H (1994) Increased formation of oxidative DNA damage, 8-hydroxydeoxyguanosine, in human livers with chronic hepatitis. Cancer Res 54(12):3171–3172

Shirakawa H, Suzuki H, Shimomura M, Kojima M, Gotohda N, Takahashi S, Nakagohri T, Konishi M, Kobayashi N, Kinoshita T, Nakatsura T (2009) Glypican-3 expression is correlated with poor prognosis in hepatocellular carcinoma. Cancer Sci 100(8):1403–1407

Shweiki D, Itin A, Soffer D, Keshet E (1992) Vascular endothelial growth factor induced by hypoxia may mediate hypoxia-initiated angiogenesis. Nature 359:843–845

Sibold S, Roh V, Keogh A, Studer P, Tiffon C, Angst E, Vorburger SA, Weimann R, Candinas D, Stroka D (2007) Hypoxia increases cytoplasmic expression of NDRG1, but is insufficient for its membrane localization in human hepatocellular carcinoma. FEBS Lett 581:989–994

Song J, Qu Z, Guo X, Zhao Q, Zhao X, Gao L, Sun K, Shen F, Wu M, Wei L (2009) Hypoxia-induced autophagy contributes to the chemoresistance of hepatocellular carcinoma cells. Autophagy 5(8):1131–1144

Suetsugu A, Nagaki M, Aoki H, Motohashi T, Kunisada T, Moriwaki H (2006) Characterization of CD133$^+$ hepatocellular carcinoma cells as cancer stem/progenitor cells. Biochem Biophys Res Commun 351(4):820–824

Tanaka H, Yamamoto M, Hashimoto N, Miyakoshi M, Tamakawa S, Yoshie M, Tokusashi Y, Yokoyama K, Yaginuma Y, Ogawa K (2006) Hypoxia-independent overexpression of hypoxia-inducible factor 1α as an early change in mouse hepatocarcinogenesis. Cancer Res 66:11263–11270

Tanaka H, Fujita N, Sugimoto R, Urawa N, Horiike S, Kobayashi Y, Iwasa M, Ma N, Kawanishi S, Watanabe S, Kaito M, Takei Y (2008) Hepatic oxidative DNA damage is associated with increased risk for hepatocellular carcinoma in chronic hepatitis C. Br J Cancer 98(3):580–586

Theise ND, Yao JL, Harada K, Hytiroglou P, Portmann B, Thung SN, Tsui W, Ohta H, Nakanuma Y (2003) Hepatic 'stem cell' malignancies in adults: four cases. Histopathology 43(3):263–271

Thiery JP, Acloque H, Huang RY, Nieto MA (2009) Epithelial-mesenchymal transitions in development and disease. Cell 139(5):871–890

Toffoli S, Michiels C (2008) Intermittent hypoxia is a key regulator of cancer cell and endothelial cell interplay in tumours. FEBS J 275:2991–3002

Tomuleasa C, Soritau O, Rus-Ciuca D, Pop T, Todea D, Mosteanu O, Pintea B, Foris V, Susman S, Kacsó G, Irimie A (2010) Isolation and characterization of hepatic cancer cells with stem-like properties from hepatocellular carcinoma. J Gastrointesin Liver Dis 19(1):61–67

Trinchieri G (2003) Interleukin-12 and the regulation of innate resistance and adaptive immunity. Nat Rev Immunol 3(2):133–146

Tsai SM, Lin SK, Lee KT, Hsiao JK, Huang JC, Wu SH, Ma H, Wu SH, Tsai LY (2009) Evaluation of redox statuses in patients with hepatitis B virus-associated hepatocellular carcinoma. Ann Clin Biochem 46(Pt 5):394–400

Uenishi T, Kubo S, Yamamoto T, Shuto T, Ogawa M, Tanaka H, Tanaka S, Kaneda K, Hirohashi K (2003) Cytokeratin 19 expression in hepatocellular carcinoma predicts early postoperative recurrence. Cancer Sci 94(10):851–857

van Malenstein H, Gevaert O, Libbrecht L, Daemen A, Allemeersch J, Nevens F, Van Cutsem E, Cassiman D, De Moor B, Verslype C, van Pelt J (2010) A seven-gene set associated with chronic hypoxia of prognostic importance in hepatocellular carcinoma. Clin Cancer Res 16(16):4278–4288

van Zijl F, Mair M, Csiszar A, Schneller D, Zulehner G, Huber H, Eferl R, Beug H, Dolznig H, Mikulits W (2009a) Hepatic tumor-stroma crosstalk guides epithelial to mesenchymal transition at the tumor edge. Oncogene 28(45):4022–4033

van Zijl F, Zulehner G, Petz M, Schneller D, Kornauth C, Hau M, Machat G, Grubinger M, Huber H, Mikulits W (2009b) Epithelial-mesenchymal transition in hepatocellular carcinoma. Future Oncol 5(8):1169–1179

Visvader JE, Lindeman GJ (2008) Cancer stem cells in solid tumours: accumulating evidence and unresolved questions. Nat Rev Cancer 8(10):755–768

Wang Y, Kato N, Hoshida Y, Yoshida H, Taniguchi H, Goto T, Moriyama M, Otsuka M, Shiina S, Shiratori Y, Ito Y, Omata M (2003) Interleukin-1β gene polymorphisms associated with hepatocellular carcinoma in hepatitis C virus infection. Hepatology 37(1):65–71

Wang F, Jing X, Li G, Wang T, Yang B, Zhu Z, Gao Y, Zhang Q, Yang Y, Wang Y, Wang P, Du Z (2012) Foxp3$^+$ regulatory T cells are associated with the natural history of chronic hepatitis B and poor prognosis of hepatocellular carcinoma. Liver Int 32(4):644–655

Wenger RH, Rolfs A, Marti HH, Bauer C, Gassmann M (1995) Hypoxia, a novel inducer of acute phase gene expression in a human hepatoma cell line. J Biol Chem 270:27865–12870

Wheelhouse NM, Chan YS, Gillies SE, Caldwell H, Ross JA, Harrison DJ, Prost S (2003) TNF-α induced DNA damage in primary murine hepatocytes. Int J Mol Med 12(6):889–894

Wong VW, Yu J, Cheng AS, Wong GL, Chan HY, Chu ES, Ng EK, Chan FK, Sung JJ, Chan HL (2009) High serum interleukin-6 level predicts future hepatocellular carcinoma development in patients with chronic hepatitis B. Int J Cancer 124(12):2766–2770

Wu XZ, Xie GR, Chen D (2007) Hypoxia and hepatocellular carcinoma: the therapeutic target for hepatocellular carcinoma. J Gastroenterol Hepatol 22:1178–1182

Wu K, Kryczek I, Chen L, Zou W, Welling TH (2009) Kupffer cell suppression of CD8$^+$ T cells in human hepatocellular carcinoma is mediated by B7-H1/programmed death-1 interactions. Cancer Res 69(20):8067–8075

Wu SD, Ma YS, Fang Y, Liu LL, Fu D, Shen XZ (2012) Role of the microenvironment in hepatocellular carcinoma development and progression. Cancer Treat Rev 38(3):218–225

Xiong YQ, Sun HC, Zhang W, Zhu XD, Zhuang PY, Zhang JB, Wang L, Wu WZ, Qin LX, Tang ZY (2009) Human hepatocellular carcinoma tumor-derived endothelial cells manifest increased angiogenesis capability and drug resistance compared with normal endothelial cells. Clin Cancer Res 15(15):4838–4846

Yamashita YI, Shimada M, Hasegawa H, Minagawa R, Rikimaru T, Hamatsu T, Tanaka S, Shirabe K, Miyazaki JI, Sugimachi K (2001) Electroporation-mediated interleukin-12 gene therapy for hepatocellular carcinoma in the mice model. Cancer Res 61(3):1005–1012

Yamashita T, Forgues M, Wang W, Kim JW, Ye Q, Jia H, Budhu A, Zanetti KA, Chen Y, Qin LX, Tang ZY, Wang XW (2008) EpCAM and alpha-fetoprotein expression defines novel prognostic subtypes of hepatocellular carcinoma. Cancer Res 68(5):1451–1461

Yamashita T, Ji J, Budhu A, Forgues M, Yang W, Wang HY, Jia H, Ye Q, Qin LX, Wauthier E, Reid LM, Minato H, Honda M, Kaneko S, Tang ZY, Wang XW (2009) EpCAM-positive hepatocellular carcinoma cells are tumor-initiating cells with stem/progenitor cell features. Gastroenterology 136(3):1012–1024

Yan W, Fu Y, Tian D, Liao J, Liu M, Wang B, Xia L, Zhu Q, Luo M (2009) PI3 kinase/Akt signaling mediates epithelial-mesenchymal transition in hypoxic hepatocellular carcinoma cells. Biochem Biophys Res Commun 382(3):631–636

Yan W, Chang Y, Liang X, Cardinal JS, Huang H, Thorne SH, Monga SP, Geller DA, Lotze MT, Tsung A (2012) High-mobility group box 1 activates caspase-1 and promotes hepatocellular carcinoma invasiveness and metastases. Hepatology 55(6):1863–1875

Yang Z, Ho DW, Ng MN, Lau NCK, Yu WC, Ngai P, Chu PWK, Lam CT, Poon RTP, Fan ST (2008) Significance of CD90$^+$ cancer stem cells in human liver cancer. Cancer Cell 13:153–166

Yang MH, Chen CL, Chau GY, Chiou SH, Su CW, Chou TY, Peng WL, Wu JC (2009) Comprehensive analysis of the independent effect of twist and snail in promoting metastasis of hepatocellular carcinoma. Hepatology 50(5):1464–1474

Yang JD, Nakamura I, Roberts LR (2011) The tumor microenvironment in hepatocellular carcinoma: current status and therapeutic targets. Semin Cancer Biol 21(1):35–43

Yin S, Li J, Hu C, Chen X, Yao M, Yan M, Jiang G, Ge C, Xie H, Wan D, Yang S, Zheng S, Gu J (2007) CD133 positive hepatocellular carcinoma cells possess high capacity for tumorigenicity. Int J Cancer 120(7):1444–1450

Yu H, Pardoll D, Jove R (2009) STATs in cancer inflammation and immunity: a leading role for STAT3. Nat Rev Cancer 9(11):798–809

Zhang Y, Yang B, Du Z, Bai T, Gao YT, Wang YJ, Lou C, Wang FM, Bai Y (2007) Aberrant methylation of SPARC in human hepatocellular carcinoma and its clinical implication. World J Gastroenterol 18(17):2043–2052

Zhu Z, Hao X, Yan M, Yao M, Ge C, Gu J, Li J (2010) Cancer stem/progenitor cells are highly enriched in CD133$^+$CD44$^+$ population in hepatocellular carcinoma. Int J Cancer 126(9):2067–2078

Chapter 5
Molecular Mechanisms of HCC

Over the years, it has become clear that various risk factors can trigger the activation of a number of cellular signaling cascades that cause the development and progression of HCC (Figs. 2.1 and 5.1). Even though the activation of these pathways is dependent upon the stimulus, several of them are simultaneously induced with a significant level of 'crosstalk' between different pathways. Similarly, the expression of a large number of microRNAs is up-regulated indicating various roles these small molecules play in the pathogenesis of liver cancer.

Signaling Pathways

Viral infection, exposure to hepatotoxic agents, and risk factors cause abrupt changes in the cellular signaling pathways and alter gene expression resulting in tumor formation. Hepatocarcinogenesis is a multistep process involving a number of factors. These include mutations that cause genetic alterations, aberrant expression of cellular proteins, inhibition of tumor suppressors, overexpression of oncogenes, and molecules that regulate these events including microRNAs and various cellular proteins (Aravalli et al. 2008). Initial studies on the pathogenesis of HCC have identified activation of a number of critical signaling pathways, as well as the mutations that activate oncogenes (β-catenin, Axin1, PI-3-kinase, Kras) and inactivate tumor suppressors (p53, Rb1, CDKN2A, IGF2R, PTEN) (Aravalli et al. 2008; van Malenstein et al. 2010). Subsequent studies using cutting edge omics-based approaches have been instrumental in the identification of potential biomarkers and molecular targets for the treatment of HCC.

© The Author(s) 2014
R.N. Aravalli, C.J. Steer, *Hepatocellular Carcinoma*, SpringerBriefs
in Cancer Research, DOI 10.1007/978-3-319-09414-4_5

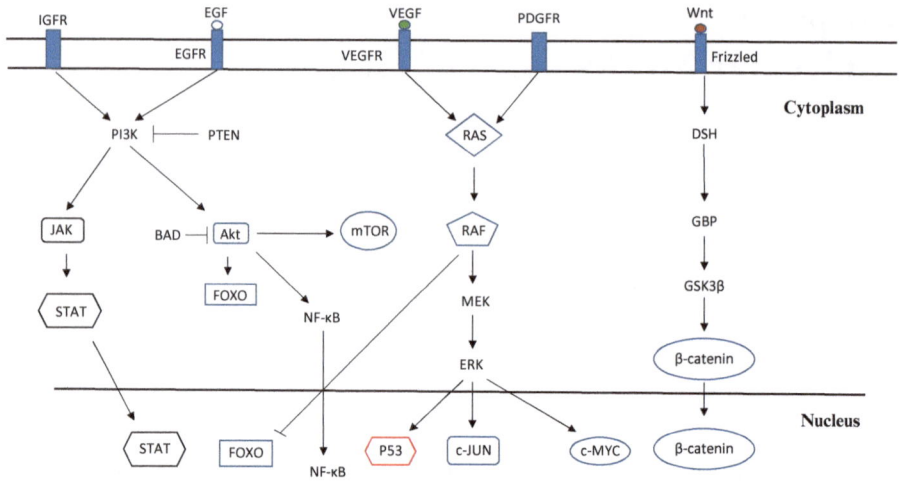

Fig. 5.1 Activation of growth factor receptor signaling pathways during HCC. Molecular targets for HCC therapy are being developed either to inhibit or activate each of these pathways as well as their associated proteins. Abbreviations: *EGF* epidermal growth factor, *EGFR* EGF receptor, *VEGF* vascular endothelial growth factor, *VEGFR* VEGF receptor, *IGFR* insulin-like growth factor 1 receptor, *PDGFR* platelet-derived growth factor receptor, *PI3K* phosphoinositide 3-kinase, *PTEN* phosphatase and tensin homolog, *BAD* Bcl-2-associated death promoter, *Akt* protein kinase B, *mTOR* mammalian target of rapamycin, *FOXO* forkhead box protein class O, *MEK* mitogen-activated protein kinase kinase, *ERK* extracellular signal-regulated kinase, *DSH* dishevelled, *GSK-3β* glycogen synthase kinase 3 beta, *GBP* GSK3 binding protein

Wnt/β-Catenin Signaling Pathway

The Wnt/β-catenin signaling pathway is highly conserved and involved in diverse cellular processes such as homeostasis, cell proliferation, differentiation, motility and apoptosis. It is deregulated in a number of cancers including HCC (Polakis 2012). Alterations of this pathway are an early oncogenic event(s) in HCC caused by HBV/HCV infections and alcoholic liver cirrhosis; and activation of Wnt signaling produces an inflammatory response (Anson et al. 2012). Inactivation of the tumor-suppressor gene adenomatous polyposis coli (APC) or mutation of the proto-oncogene β-catenin also frequently occurs leading to the activation of Wnt signaling (de la Coste et al. 1998). Liver-specific disruption of APC gene in mice was shown to cause the activation of Wnt pathway leading to the development of HCC (Colnot et al. 2005). Gain-of-function mutations in the gene encoding β-catenin are common genetic modifications in human HCC; and in primary hepatocytes, β-catenin signaling leads to activation of the transcription factor nuclear factor-κB (NF-κB) pathway (Anson et al. 2012). NF-κB is a master regulator of inflammation and cell death in the development of hepatocellular injury, liver fibrosis and HCC (Elsharkawy and Mann 2007; Luedde and Schwabe 2011), and therefore, is a therapeutic target for liver cancer. Upregulation of frizzled-7 and dephosphorylation of β-catenin are frequently observed in HCC (Merle et al. 2005). Mutations in β-catenin arise in HCC patients with increased exposure to either HCV infection or aflatoxin (Lee et al. 2006).

In liver, members of the glutathione S-transferase (GST) family are involved in the detoxification of xenobiotic compounds and biotransformation of endogenous substrates, and play defensive roles against electrophilic compounds and oxidative stress (Hayes et al. 2005). In mouse hepatocytes, the Wnt/β-catenin pathway controls the expression of GSTs suggesting that it can regulate expression of both phase I and phase II drug-metabolizing enzymes (Giera et al. 2010). Altered expression of GSTs due to abnormal Wnt signaling has been observed in human and rodent HCC (Stahl et al. 2005; Hailfinger et al. 2006). CD24 overexpression in HCC leads to activation of Wnt/β-catenin pathway and is associated with increased invasiveness and metastatic potential (Yang et al. 2009). Collectively, these findings imply that targeted inactivation of the Wnt/β-catenin pathway is a potential therapeutic target for HCC.

p53 Pathway

Normally, cellular levels of the tumor suppressor p53 are low, but rapidly upregulated in response to intracellular and extracellular stress signals. In about 50 % of all human tumors, the tumor suppressor gene TP53 is inactivated by a single point mutation, and in others, the p53 protein is expressed at normal levels but the signaling pathways that lead to cell cycle arrest and apoptosis are defective (Stegh 2012). Mutations that commonly alter p53 function include G:C to T:A transversions at codon 249, and C:G to A:T and C:G to T:A transversions at codon 250. A number of studies have reported p53 mutations and its inactivation in HCC (Aravalli et al. 2008). For instance, a recent meta-analysis has shown that HCC patients with upregulated mutant p53 expression have shorter overall survival than those with wild type p53 (Liu et al. 2012). Interestingly, the R249S mutation accounts for almost 90 % of all mutations in p53 seen in AFB_1-related HCC as well as in HCC cell lines (Gouas et al. 2010). Importantly, the expression of this mutant p53 in the plasma might serve as a potential biomarker for HCC.

Rb Pathway

The retinoblastoma protein pRb is a major cellular tumor suppressor and it prevents cancer development by controlling cell cycle progression through the inhibition of E2F transcription factor family of proteins (Goodrich 2006). Cyclin-dependent kinases (CDKs) phosphorylate and activate pRb to induce G1/S cell cycle transition. Multiple CDKs can phosphorylate any of 16 putative phosphorylation sites on pRb with varying degrees of specificity. Furthermore, a strong correlation has been

observed between the loss of pRb activity and lack of functional p53 in the early
studies on human cancers, including HCC (Aravalli et al. 2008). In about 90 %
of HCC cases, CDK inhibitors p16INK4A, p21(WAF1/CIP1) and p27Kip1 are
inactivated and a change in their expression contributes to carcinogenesis both
during the early stages of development and during disease progression (Harbour
and Dean 2000). Interestingly, loss of Rb function does not provide proliferative
advantage to Myc-expressing HCC cells (Saddic et al. 2011). Instead, it caused an
increase in polyploidy in mature hepatocytes before the development of tumors. In
a model of HCC, it has been demonstrated that the inactivation of three pRb family
members (Rb, p107 and p130) leads to the activation of Notch signaling, which, in
turn, inhibits HCC (Viatour et al. 2011). In light of these findings, pRb constitutes
an additional and potentially effective therapeutic target to treat HCC.

Ras Pathway

Human ras proteins H-Ras, N-Ras, K-Ras4A and K-Ras4B are small GTP-binding
proteins that influence cell growth, differentiation and apoptosis (Röring and
Brummer 2012). Ras interacts with a downstream serine/threonine kinase Raf-1
leading to its activation and downstream signaling, that includes activation of
MAPK kinases (MKKs) MEK1 and MEK2 to regulate proliferation and apoptosis.
Activation of Ras and proteins of its pathway such as p21 was reported in HCC
(Calvisi et al. 2006; Aravalli et al. 2008). The strategies of inhibiting several kinases
and suppressing Ras expression using antisense RNA were successfully applied in
cell lines and animal models (Liao et al. 2000). It has been suggested that the Ras
pathway is important in HCC of rodents but not for humans based upon the observed
low mutation rate of Ras (Grisham 1997). However, in a recent study, it was reported
that RASSF1A and NORE1A, members of the RASSF family of Ras inhibitors,
were inactivated in human HCC demonstrating a potential role for Ras pathway
in liver cancer (Calvisi et al. 2006). Single point mutations in codon 13 of H-ras,
codon 12 of N-ras and codon 61 of K-ras were also observed in HCC (Aravalli et al.
2008). Furthermore, the expression levels of Spred protein (Sprouty-related protein
with Ena/vasodilator-stimulated phosphoprotein homology-1 domain), an inhibitor
of the Ras/Raf-1/ERK pathway were also found to be deregulated in HCC (Yoshida
et al. 2006).

MAPK Pathway

The mitogen-activated protein (MAP) kinase family of proteins is implicated
in diverse cellular processes such as cell survival, differentiation, adhesion, and
proliferation (del Barco Barrantes and Nebreda 2012). Members of this family
include extracellular signal-regulated kinase protein homologs 1 and 2 (ERK1/2),

big MAPK-1 (BMK-1/ERK5), c-Jun N-terminal kinase homologs 1, 2 and 3 (JNK1/2/3), stress-activated protein kinase 2 (SAPK-2) homologs α, β, and δ (p38α/β/δ), and ERK6, also known as p38γ (Pimienta and Pascual 2007; Davies and Tournier 2012). Several studies have shown that MAPK signaling is a factor in HCC, and proteins of HBV, HCV and HEV modulate MAPKs by targeting multiple steps along the signaling pathway(s) (Pleschka 2008). For instance, HCV E2 protein activates MAPK pathway in HuH7 cells and promotes cell proliferation (Zhao et al. 2005). A direct correlation of MAPK-ERK pathway in HCC was identified in another study when Spread was overexpressed in cell lines. In these studies, Spread inhibited ERK activation, both *in vivo* and *in vitro*, which resulted in reduced proliferation of cancer cells and low secretion of MMP-2 and MMP-9 (Yoshida et al. 2006).

In addition to their role in the development of liver fibrosis, activated HSCs promote HCC cell proliferation. The conditioned media collected from HSCs induced proliferation and migration of HCC cells cultured in monolayers. Moreover, HSCs promoted HCC growth in a 3-dimensional spheroid co-culture system and diminished the extent of central necrosis through the activation of NF-κB and extracellular-regulated kinase (ERK) pathways (Amann et al. 2009). Consistent with these findings, simultaneous *in vivo* implantation of HSCs and HCC cells into nude mice promoted tumor growth and invasiveness, and inhibited necrosis. PDGF, TGF-β1, MMP-9, JNK, insulin-like growth factor binding protein 5, cathepsins B and D, hepatitis B virus X protein, and HCV nonstructural proteins are all potent inducers of stellate cell activation, proliferation and collagen production, and therefore enhance liver fibrosis and carcinogenesis (Campbell et al. 2005; Schulze-Krebs et al. 2005; Martin-Vilchez et al. 2008; Moles et al. 2009; Kluwe et al. 2011). In contrast, adiponectin suppressed HSC activation and Th1 immune responses in tumor.

JAK/STAT Pathway

Signal transducers and activators of transcription (STATs) are a family of tran-scription factors regulated by a variety of cytokines, hormones and growth factors (Harrison 2012). Their activation occurs through tyrosine phosphorylation by JAKs. STATs stimulate the transcription of suppressors of cytokine signaling (SOCS) genes and their proteins, and in turn, bind phosphorylated JAKs to inhibit the pathway and prevent overactivation of cytokine-stimulated cells (Croker et al. 2008). Thus, SOCS are part of the negative feedback loop in the JAK/STAT circuitry. JAK stimulation of STATs activates cell proliferation, migration, differentiation and apoptosis, and deregulation of inhibitors leads to human diseases, including cancer. It has been reported that inactivation of SOCS-1 and SSI-1 and activation of JAK/STAT pathway occur in HCC (Calvisi et al. 2006). Thus, the JAK/STAT pathway also plays an important role in liver cancer. Moreover, it has been shown that IL-6 is a link between obesity and HCC as increased expression of IL-6 and

TNF in obese mice leads to the activation of the IL-6 signaling pathway via the downstream STAT3 and ERK pathways, thus promoting tumorigenesis in the liver (Park et al. 2007).

Heat Shock Proteins

Heat shock proteins (HSPs) are critical players in cellular stress response via phosphorylation/dephosphorylation. In a recent study conducted with 48 clinical specimens, HCC progression was found to be associated with a decrease in serine phosphorylation of HSP27 (Yasuda et al. 2005). In another study with 146 clinical specimens, several members of the HSP family were found to be associated with HCC (Luk et al. 2006), suggesting that HSPs may be key players in tumor progression. HSP20 is typically downregulated in HCC; and in a recent study, overexpression of HSP20 in HCC cell lines caused inhibition of MAPK and Akt pathways and suppression of cell proliferation (Matsushima-Nishiwaki et al. 2011). The effect of oxidative stress during liver carcinogenesis, most notably in hemochromatosis and Wilson disease, was shown to be associated with p53 mutations (Hussain et al. 2000). Under such conditions, oxidative stress resulted in an elevated level of inducible nitric oxide synthase (iNOS) expression and development of cirrhosis with a 200-fold risk for HCC. VEGF and FGF play important roles in HCC development (Yoshiji et al. 2005; Huang et al. 2006). In a recent study, it was shown that the use of inhibitors of EGFR and TGF-β prevented the development of HCC in the rat liver demonstrating the potentially harmful nature of these growth factors (Schiffer et al. 2005).

Others

Absence of apoptotic cell death is one of the hallmarks of liver cancer. By suppressing the expression of an anti-apoptotic myeloid cell leukemia-1 protein (Mcl-1) using RNA interference, it was demonstrated that induction of apoptosis could be achieved in HCC cancer cells *in vitro* (Schulze-Bergkamen et al. 2006). Additional pathways of physiological processes involving alcohol metabolism, cellular transport, and ubiquitination may all play a role in regulating the development and progression of liver cancer. Furthermore, it was reported recently that inflammation is inherently associated with cancer and a number of cytokines are involved in promoting HCC development and progression, especially during viral infection (Budhu et al. 2006). In particular, Th2 cytokines are induced and Th1 cytokines decreased in metastases. Therefore, modulating the expression of cytokines and the use of inhibitors of inflammatory cytokines might be critical in modulating HCC progression.

MicroRNAs

MicroRNAs (miRNAs) are transcribed by RNA polymerase II to form primary transcripts (pri-miRNAs), capped with 7-methylguanosine and polyadenylated. These pri-miRNAs are processed in the nucleus by the RNase III enzyme Drosha and its cofactor Pasha/DGCR8 into ∼60–70 nucloetide precursors (pre-miRNAs), which form an imperfect stem-loop structure and are exported into the cytoplasm by the RAN GTP-dependent transporter Exportin 5. RNase III enzyme Dicer then processes pre-miRNAs into mature miRNAs, which are subsequently loaded onto the RNA-induced silencing complex (RISC). The miRNA/RISC complex then binds to the 3′UTR of target mRNAs and downregulates their expression (van Kouwenhove et al. 2011; Broderick and Zamore 2011). Thus, miRNAs modulate various cellular signaling pathways involved in cell growth, proliferation, motility, and survival. miRNAs are also subjected to regulation by epigenetic mechanisms, and mutations in their promoter and coding regions were shown to contribute to tumorigenesis (Datta et al. 2008; Xu et al. 2008, 2011; Furuta et al. 2010; Wang et al. 2012).

During the past decade, it has become well established that specific miRNAs modulate various cellular processes in the liver, and that their aberrant expression correlates with the severity and prognosis of HCC (Table 5.1) (Calin et al. 2004; Murakami et al. 2006; Gramantieri et al. 2008; Ura et al. 2009; Huang et al. 2009). For example, in one study, the expression of miR-199a, miR-92, miR-106a, miR-222, miR-17-5p, miR-18, and miR-20 correlated with the degree of tumor differentiation, suggesting the involvement of these miRNAs in disease progression (Murakami et al. 2006). In a separate study, miRNAs associated with viral infection and the stage of liver disease were identified by comparing miRNA expression profiles of patients with HBV-related HCC with those from HCV-related HCC (Ura et al. 2009). Further analysis of HBV-associated miRNAs revealed that their targets included genes involved in pathways related to cell death, DNA damage, recombination, and signal transduction. Additional targets included genes involved in immune response, antigen presentation, cell cycle, proteasome, and lipid metabolism (Ura et al. 2009). It appears that HBV and HCV induce different sets of miRNAs during infection and the potential for unique biomarkers during disease progression.

miRNA-122, the most abundant liver-specific miRNA modulates hepatic lipid metabolism (Lindow and Kauppinen 2012) and is often downregulated in human HCC (Kutay et al. 2006; Jopling 2012). Loss of its expression correlates with loss of mitochondrial metabolic function, and is detrimental to sustaining critical liver function, thereby contributing to morbidity and mortality (Burchard et al. 2010). In the HBV-related HCC cell line HepG2.2.15, miR-122 inhibited viral replication by targeting NDRG3, a member of the N-myc downstream-regulated gene family (Fan et al. 2011), suggesting that both miRNA-122 and NDRG3 are viable therapeutic targets for HBV-related HCC. During HCV infection, however, miR-122 binds directly to two sites in the 5′ non-coding region of HCV genome and

Table 5.1 MicroRNAs aberrantly expressed in HCC tumor tissues and cell lines, and their identified role(s) in various cellular mechanism(s)

Sample type	Method	MicroRNA	Cellular mechanism/target
Tumor tissues	Microarray	miR-199a, miR-92, miR-106a	Tumor progression
	Q-PCR	miR-222, miR-17-5p, miR-18, miR-20	
	Microarray	miR-122, let-7a, miR-21, miR-23	Tumorigenesis
	Microarray	miR-199-a-5p, miR-223	Cell cycle inhibition
	Q-PCR	miR-34a	Apoptosis, cell cycle
	Western blot		Arrest, senescence, c-Met
	Microarray	miR-101	Tumor suppression
	Northern blot		Apoptosis, Mcl-1
	Western blot		
	Q-PCR	miR-21	PTEN
	Q-PCR	miR-221	CDKN1C/p27
	Northern blot		CDKN1C/p57
	Microarray	miR-122, miR-145, let-7 family	Cyclin G1
	Northern blot		
	Q-PCR		
HCC cell lines	Q-PCR	miR-34a	Apoptosis, cell cycle
	Western blot		Arrest, senescence, c-Met
	Microarray	miR-101	Tumor suppression
	Northern blot		Apoptosis, Mcl-1
	Western blot		
	Northern blot	miR-17-92, miR-21	Cell proliferation
	Q-PCR		Apoptosis
	Microarray	miR-122, miR-145, let-7 family	Cyclin G1
	Q-PCR		
	Northern blot		
	Western blot	miR-122	Migration, invasion
	Soft agar assay		Anchorage-dependence
	Q-PCR		Angiogenesis, Metastasis, ADAM17, NDRG3, Cyclin G1
Rat model of hepatoma	Northern blot	miR-130, miR-190, miR-19-72 family	Morbidity, mortality, loss of mitochondrial function

Modified from Aravalli (2013) with permission, MDPI AG

positively regulates the viral life cycle (Jopling 2012). Thus, inhibition of miR-122 presents an attractive treatment option for HCV infection (Janssen et al. 2013).

In another study, examination of miRNAs in HCC with cirrhotic background revealed that members of the let-7 family, miR-221, and mir-145 were downregulated (Gramantieri et al. 2007). In these tissues and HCC cell lines, miR-122a was also downregulated, and its target gene product cyclin G1 was highly expressed and promoting the growth of cancer cells. However, when miR-122 expression was restored, it significantly reduced *in vitro* migration, invasion, and anchorage-independent growth of Mhalavu and SK-HEP-1 cells. It also reduced *in vivo* tumorigenesis, angiogenesis, and intrahepatic metastasis in an orthotopic liver cancer model by ADAM17, a protein involved in metastasis (Tsai et al. 2009). Interestingly, miR-122 also inhibited tumorigenic properties of HCC, and sensitized tumor cells to Sorafenib (Bai et al. 2009). Collectively, these studies demonstrate that miR-122 can function both as a positive and negative regulator in liver cancer. Other studies have shown that miRNAs associated with cell cycle inhibition (miR-34a, miR-101, miR199-a-5p, and miR-223) were downregulated in HCC (Datta et al. 2008; Ladeiro et al. 2008; Li et al. 2008, 2009; Wong et al. 2008; Yang et al. 2008; Su et al. 2009), and those involved in cell proliferation and inhibition of apoptosis (miR-17-92 polycistron, miR-21, miR-34a, miR-96, miR-221, and miR-224) were upregulated (Meng et al. 2007; Connolly et al. 2008; Fornari et al. 2008; Wang et al. 2008). Furthermore, miRNA-221 is associated with tumor multifocality (Gramantieri et al. 2009). As demonstrated in these examples, aberrant miRNA expression leads to the dysregulation of critical cellular mechanisms, and activation of tumorigenic pathways involved in tumor differentiation, diagnosis, staging, progression, prognosis and response to therapy.

References

Amann T, Bataille F, Spruss T, Muhlbauer M, Gabele E, Scholmerich J, Kiefer P, Bosserhoff AK, Hellerbrand C (2009) Activated hepatic stellate cells promote tumorigenicity of hepatocellular carcinoma. Cancer Sci 100(4):646–653

Anson M, Crain-Denoyelle AM, Baud V, Chereau F, Gougelet A, Terris B, Yamagoe S, Colnot S, Viguier M, Perret C, Couty JP (2012) Oncogenic β-catenin triggers an inflammatory response that determines the aggressiveness of hepatocellular carcinoma in mice. J Clin Invest 122(2):586–599

Aravalli RN (2013) Development of microRNA therapeutics for hepatocellular carcinoma. Diagnostics 3:170–191

Aravalli RN, Steer CJ, Cressman EN (2008) Molecular mechanisms of hepatocellular carcinoma. Hepatology 48(6):1049–1053

Bai S, Nasser MW, Wang B, Hsu SH, Datta J, Kutay H, Yadav A, Nuovo G, Kumar P, Ghoshal K (2009) MicroRNA-122 inhibits tumorigenic properties of hepatocellular carcinoma cells and sensitizes these cells to sorafenib. J Biol Chem 284(46):32015–32027

Broderick JA, Zamore PD (2011) MicroRNA therapeutics. Gene Ther 18(12):1104–1110

Budhu A, Forgues M, Ye QH, Jia HL, He P, Zanetti KA, Kammula US, Chen Y, Qin LX, Tang ZY, Wang XW (2006) Prediction of venous metastases, recurrence and prognosis in hepatocellular carcinoma based on a unique immune response signature of the liver microenvironment. Cancer Cell 10(2):99–111

Burchard J, Zhang C, Liu AM, Poon RT, Lee NP, Wong KF, Sham PC, Lam BY, Ferguson MD, Tokiwa G, Smith R, Leeson B, Beard R, Lamb JR, Lim L, Mao M, Dai H, Luk JM (2010) microRNA-122 as a regulator of mitochondrial metabolic gene network in hepatocellular carcinoma. Mol Syst Biol 6:402

Calin GA, Sevignani C, Dumitru CD, Hyslop T, Noch E, Yendamuri S, Shimizu M, Rattan S, Bullrich F, Negrini M, Croce CM (2004) Human microRNA genes are frequently located at fragile sites and genomic regions involved in cancers. Proc Natl Acad Sci U S A 101(9): 2999–3004

Calvisi D, Ladu S, Gorden A, Farina M, Conner EA, Lee JS, Factor VM, Thorgeirsson SS (2006) Ubiquitous activation of Ras and Jak/Stat pathways in human HCC. Gastroenterology 130(4):1117–1128

Campbell JS, Hughes SD, Gilbertson DG, Palmer TE, Holdren MS, Haran AC, Odell MM, Bauer RL, Ren HP, Haugen HS, Yeh MM, Fausto N (2005) Platelet-derived growth factor C induces liver fibrosis, steatosis, and hepatocellular carcinoma. Proc Natl Acad Sci U S A 102(9): 3389–3394

Colnot S, Decaens T, Niwa-Kawakita M, Godard C, Hamard G, Kahn A, Giovannini M, Perret C (2005) Liver-targeted disruption of *Apc* in mice activates beta-catenin signaling and leads to hepatocellular carcinomas. Proc Natl Acad Sci U S A 101(49):17216–17221

Connolly E, Melegari M, Landgraf P, Tchaikovskaya T, Tennant BC, Slagle BL, Rogler LE, Zavolan M, Tuschl T, Rogler CE (2008) Elevated expression of the miR-17–92 polycistron and miR-21 in hepadnavirus-associated hepatocellular carcinoma contributes to the malignant phenotype. Am J Pathol 173(3):856–864

Croker BA, Kiu H, Nicholson SE (2008) SOCS regulation of the JAK/STAT signalling pathway. Semin Cell Dev Biol 19(4):412–422

Datta J, Kutay H, Nasser MW, Nuovo GJ, Wang B, Majumder S, Liu CG, Volinia S, Croce CM, Schmittgen TD, Ghoshal K, Jacob ST (2008) Methylation mediated silencing of MicroRNA-1 gene and its role in hepatocellular carcinogenesis. Cancer Res 68(13):5049–5058

Davies C, Tournier C (2012) Exploring the function of the JNK (c-Jun N-terminal kinase) signalling pathway in physiological and pathological processes to design novel therapeutic strategies. Biochem Soc Trans 40(1):85–89

de La Coste A, Romagnolo B, Billuart P, Renard CA, Buendia MA, Soubrane O, Fabre M, Chelly J, Beldjord C, Kahn A, Perret C (1998) Somatic mutations of the beta-catenin gene are frequent in mouse and human hepatocellular carcinomas. Proc Natl Acad Sci U S A 95(15):8847–8851

del Barco Barrantes I, Nebreda AR (2012) Roles of p38 MAPKs in invasion and metastasis. Biochem Soc Trans 40(1):79–84

Elsharkawy AM, Mann DA (2007) Nuclear factor-kB and the hepatic inflammation-fibrosis-cancer axis. Hepatology 46(2):590–597

Fan CG, Wang CM, Tian C, Wang Y, Li L, Sun WS, Li RF, Liu YG (2011) miR-122 inhibits viral replication and cell proliferation in hepatitis B virus-related hepatocellular carcinoma and targets NDRG3. Oncol Rep 26(5):1281–1286

Fornari F, Gramantieri L, Ferracin M, Veronese A, Sabbioni S, Calin GA, Grazi GL, Giovannini C, Croce CM, Bolondi L, Negrini M (2008) MiR-221 controls CDKN1C/p57 and CDKN1B/p27 expression in human hepatocellular carcinoma. Oncogene 27(43):5651–5661

Furuta M, Kozaki KI, Tanaka S, Arii S, Imoto I, Inazawa J (2010) miR-124 and miR-203 are epigenetically silenced tumor-suppressive microRNAs in hepatocellular carcinoma. Carcinogenesis 31(5):766–776

Giera S, Braeuning A, Köhle C, Bursch W, Metzger U, Buchmann A, Schwarz M (2010) Wnt/β-catenin signaling activates and determines hepatic zonal expression of glutathione S-transferases in mouse liver. Toxicol Sci 115(1):22–33

Goodrich DW (2006) The retinoblastoma tumor-suppressor gene, the exception that proves the rule. Oncogene 25(38):5233–5243

Gouas DA, Shi H, Hautefeuille AH, Ortiz-Cuaran SL, Legros PC, Szymanska KJ, Galy O, Egevad LA, Abedi-Ardekani B, Wiman KG, Hantz O, Caron de Fromentel C, Chemin IA, Hainaut PL (2010) Effects of the TP53 p.R249S mutant on proliferation and clonogenic properties in human hepatocellular carcinoma cell lines: interaction with hepatitis B virus X protein. Carcinogenesis 31(8):1475–1482

Gramantieri L, Ferracin M, Fornari F, Veronese A, Sabbioni S, Liu CG, Calin GA, Giovannini C, Ferrazzi E, Grazi GL, Croce CM, Bolondi L, Negrini M (2007) Cyclin G1 is a target of miR-122a, a microRNA frequently down-regulated in human hepatocellular carcinoma. Cancer Res 67(13):6092–6099

Gramantieri L, Fornari F, Callegari E, Sabbioni S, Lanza G, Croce CM, Bolondi L, Negrini M (2008) MicroRNA involvement in hepatocellular carcinoma. J Cell Mol Med 12(6A): 2189–2204

Gramantieri L, Fornari F, Ferracin M, Veronese A, Sabbioni S, Calin GA, Grazi GL, Croce CM, Bolondi L, Negrini M (2009) MicroRNA-221 targets Bmf in hepatocellular carcinoma and correlates with tumor multifocality. Clin Cancer Res 15(16):5073–5081

Grisham JW (1997) Interspecies comparison of liver carcinogenesis: implications for cancer risk assessment. Carcinogenesis 18(1):59–81

Hailfinger S, Jaworski M, Braeuning A, Buchmann A, Schwarz M (2006) Zonal gene expression in murine liver: lessons from tumors. Hepatology 43(3):407–414

Harbour JW, Dean DC (2000) The Rb/E2F pathway: expanding roles and emerging paradigms. Genes Dev 14(19):2393–2409

Harrison DA (2012) The Jak/STAT pathway. Cold Spring Harb Perspect Biol 4(3):pii: a011205

Hayes JD, Flanagan JU, Jowsey IR (2005) Glutathione transferases. Annu Rev Pharmacol Toxicol 45:51–88

Huang X, Yu C, Jin C, Kobayashi M, Bowles CA, Wang F, McKeehan WL (2006) Ectopic activity of fibroblast growth factor receptor 1 in hepatocytes accelerates hepatocarcinogenesis by driving proliferation and vascular endothelial growth factor-induced angiogenesis. Cancer Res 66(3):1481–1490

Huang XH, Wang Q, Chen JS, Fu XH, Chen XL, Chen LZ, Li W, Bi J, Zhang LJ, Fu Q, Zeng WT, Cao LQ, Tan HX, Su Q (2009) Bead-based microarray analysis of microRNA expression in hepatocellular carcinoma: miR-338 is downregulated. Hepatol Res 39(8):786–794

Hussain SP, Raja K, Amstad PA, Sawyer M, Trudel LJ, Wogan GN, Hofseth LJ, Shields PG, Billiar TR, Trautwein C, Hohler T, Galle PR, Phillips DH, Markin R, Marrogi AJ, Harris CC (2000) Increased p53 mutation load in nontumorous human liver of Wilson disease and hemochromatosis: oxyradical overload diseases. Proc Natl Acad Sci U S A 97(23): 12770–12775

Janssen HL, Reesink HW, Lawitz EJ, Zeuzem S, Rodriguez-Torres M, Patel K, van der Meer AJ, Patick AK, Chen A, Zhou Y, Persson R, King BD, Kauppinen S, Levin AA, Hodges MR (2013) Treatment of HCV infection by targeting microRNA. N Engl J Med 368(18):1685–1694

Jopling C (2012) Liver-specific microRNA-122: biogenesis and function. RNA Biol 9(2):137–142

Kluwe J, Wongsiriroj N, Troeger JS, Gwak GY, Dapito DH, Pradere JP, Jiang H, Siddiqui M, Piantedosi R, O'Byrne SM, Blaner WS, Schwabe RF (2011) Absence of hepatic stellate cell retinoid lipid droplets does not enhance hepatic fibrosis but decreases hepatic carcinogenesis. Gut 60(9):1260–1268

Kutay H, Bai S, Datta J, Motiwala T, Pogribny I, Frankel W, Jacob ST, Ghoshal K (2006) Downregulation of miR-122 in the rodent and human hepatocellular carcinomas. J Cell Biochem 99(3):671–678

Ladeiro Y, Couchy G, Balabaud C, Bioulac-Sage P, Pelletier L, Rebouissou S, Zucman-Rossi J (2008) MicroRNA profiling in hepatocellular tumors is associated with clinical features and oncogene/tumor suppressor gene mutations. Hepatology 47(6):1955–1963

Lee HC, Kim M, Wands JR (2006) Wnt/Frizzled signaling in hepatocellular carcinoma. Front Biosci 11:1901–1915

Li W, Xie L, He X, Li J, Tu K, Wei L, Wu J, Guo Y, Ma X, Zhang P, Pan Z, Hu X, Zhao Y, Xie H, Jiang G, Chen T, Wang J, Zheng S, Cheng J, Wan D, Yang S, Li Y, Gu J (2008) Diagnostic and prognostic implications of microRNAs in human hepatocellular carcinoma. Int J Cancer 123(7):1616–1622

Li N, Fu H, Tie Y, Hu Z, Kong W, Wu Y, Zheng X (2009) miR-34a inhibits migration and invasion by down-regulation of c-Met expression in human hepatocellular carcinoma cells. Cancer Lett 275(1):44–53

Liao Y, Tang ZY, Ye SL, Liu KD, Sun FX, Huang Z (2000) Modulation of apoptosis, tumorigenesity and metastatic potential with antisense H-ras oligodeoxynucleotides in a high metastatic tumor model of hepatoma: LCI-D20. Hepatogastroenterology 47(32):365–370

Lindow M, Kauppinen S (2012) Discovering the first microRNA-targeted drug. J Cell Biol 199(3):407–412

Liu J, Ma Q, Zhang M, Wang X, Zhang D, Li W, Wang F, Wu E (2012) Alterations of TP53 are associated with a poor outcome for patients with hepatocellular carcinoma: evidence from a systematic review and meta-analysis. Eur J Cancer 48(15):2328–2338

Luedde T, Schwabe RF (2011) NF-kB in the liver – linking injury, fibrosis and hepatocellular carcinoma. Nat Rev Gastroenterol Hepatol 8(2):108–118

Luk JM, Lam CT, Siu AF, Lam BY, Ng IO, Hu MY, Che CM, Fan ST (2006) Proteomic profiling of hepatocellular carcinoma in Chinese cohort reveals heat-shock proteins (Hsp27, Hsp70, GRP78) up-regulation and their associated prognostic values. Proteomics 6(3):1049–1057

Martin-Vilchez S, Sanz-Cameno P, Rodriguez-Munoz Y, Majano PL, Molina-Jimenez F, Lopez-Cabrera M, Moreno-Otero R, Lara-Pezzi E (2008) The hepatitis B virus X protein induces paracrine activation of human hepatic stellate cells. Hepatology 47(6):1872–1883

Matsushima-Nishiwaki R, Adachi S, Yoshioka T, Yasuda E, Yamagishi Y, Matsuura J, Muko M, Iwamura R, Noda T, Toyoda H, Kaneoka Y, Okano Y, Kumada T, Kozawa O (2011) Suppression by heat shock protein 20 of hepatocellular carcinoma cell proliferation via inhibition of the mitogen-activated protein kinases and AKT pathways. J Cell Biochem 112(11):3430–3439

Meng F, Henson R, Wehbe-Janek H, Ghoshal K, Jacob ST, Patel T (2007) MicroRNA-21 regulates expression of the PTEN tumor suppressor gene in human hepatocellular cancer. Gastroenterology 133(2):647–658

Merle P, Kim M, Herrmann M, Gupte A, Lefrançois L, Califano S, Trépo C, Tanaka S, Vitvitski L, de la Monte S, Wands JR (2005) Oncogenic role of the frizzled-7/beta-catenin pathway in hepatocellular carcinoma. J Hepatol 43(5):854–862

Moles A, Tarrats N, Fernandez-Checa JC, Mari M (2009) Cathepsins B and D drive hepatic stellate cell proliferation and promote their fibrogenic potential. Hepatology 49(4):1297–1307

Murakami Y, Yasuda T, Saigo K, Urashima T, Toyoda H, Okanoue T, Shimotohno K (2006) Comprehensive analysis of microRNA expression patterns in hepatocellular carcinoma and non-tumor tissue. Oncogene 25(17):2537–2545

Park EJ, Lee JH, Yu GY, He G, Ali SR, Holzer RG, Osterreicher CH, Takahashi H, Karin M (2007) Dietary and genetic obesity promote liver inflammation and tumorigenesis by enhancing IL-6 and TNF expression. Cell 140(2):197–208

Pimienta G, Pascual J (2007) Canonical and alternative MAPK signaling. Cell Cycle 6(21): 2628–2632

Pleschka S (2008) RNA viruses and the mitogenic Raf/MEK/ERK signal transduction cascade. Biol Chem 389(10):1273–1282

Polakis P (2012) Wnt signaling in cancer. Cold Spring Harb Perspect Biol 4(5):pii: a008052

Röring M, Brummer T (2012) Aberrant B-Raf signaling in human cancer – 10 years from bench to bedside. Crit Rev Oncog 17(1):97–121

Saddic LA, Wirt S, Vogel H, Felsher DW, Sage J (2011) Functional interactions between retinoblastoma and c-MYC in a mouse model of hepatocellular carcinoma. PLoS One 6(5):e19758

Schiffer EHC, Cacheux W, Wendum D, Desbois-Mouthon C, Rey C, Clergue F, Poupon R, Barbu V, Rosmorduc O (2005) Gefitinib, an EGFR inhibitor, prevents hepatocellular carcinoma development in the rat liver with cirrhosis. Hepatology 41(2):307–314

Schulze-Bergkamen H, Fleischer B, Schuchmann M, Weber A, Weinmann A, Krammer PH, Galle PR (2006) Suppression of Mcl-1 via RNA interference sensitizes human hepatocellular carcinoma cells towards apoptosis induction. BMC Cancer 6:232

Schulze-Krebs A, Preimel D, Popov Y, Bartenschlager R, Lohmann V, Pinzani M (2005) Hepatitis C virus-replicating hepatocytes induce fibrogenic activation of hepatic stellate cells. Gastroenterology 129(1):246–258

Stahl S, Ittrich C, Marx-Stoelting P, Köhle C, Altug-Teber O, Riess O, Bonin M, Jobst J, Kaiser S, Buchmann A, Schwarz M (2005) Genotype-phenotype relationships in hepatocellular tumors from mice and man. Hepatology 42(2):353–361

Stegh AH (2012) Targeting the p53 signaling pathway in cancer therapy – the promises, challenges and perils. Expert Opin Ther Targets 16(1):67–83

Su H, Yang JR, Xu T, Huang J, Xu L, Yuan Y, Zhuang SM (2009) MicroRNA-101, down-regulated in hepatocellular carcinoma, promotes apoptosis and suppresses tumorigenicity. Cancer Res 69(3):1135–1142

Tsai WC, Hsu PW, Lai TC, Chau GY, Lin CW, Chen CM, Lin CD, Liao YL, Wang JL, Chau YP, Hsu MT, Hsiao M, Huang HD, Tsou AP (2009) MicroRNA-122, a tumor suppressor microRNA that regulates intrahepatic metastasis of hepatocellular carcinoma. Hepatology 49(5): 1571–1582

Ura S, Honda M, Yamashita T, Ueda T, Takatori H, Nishino R, Sunakozaka H, Sakai Y, Horimoto K, Kaneko S (2009) Differential microRNA expression between hepatitis B and hepatitis C leading disease progression to hepatocellular carcinoma. Hepatology 49(4):1098–1112

van Kouwenhove M, Kedde M, Agami R (2011) MicroRNA regulation by RNA-binding proteins and its implications for cancer. Nat Rev Cancer 11(9):644–656

van Malenstein H, Gevaert O, Libbrecht L, Daemen A, Allemeersch J, Nevens F, Van Cutsem E, Cassiman D, De Moor B, Verslype C, van Pelt J (2010) A seven-gene set associated with chronic hypoxia of prognostic importance in hepatocellular carcinoma. Clin Cancer Res 16(16):4278–4288

Viatour P, Ehmer U, Saddic LA, Dorrell C, Andersen JB, Lin C, Zmoos AF, Mazur PK, Schaffer BE, Ostermeier A, Vogel H, Sylvester KG, Thorgeirsson SS, Grompe M, Sage J (2011) Notch signaling inhibits hepatocellular carcinoma following inactivation of the RB pathway. J Exp Med 208(10):1963–1976

Wang Y, Lee AT, Ma JZ, Wang J, Ren J, Yang Y, Tantoso E, Li KB, Ooi LL, Tan P, Lee CG (2008) Profiling microRNA expression in hepatocellular carcinoma reveals microRNA-224 up-regulation and apoptosis inhibitor-5 as a microRNA-224-specific target. J Biol Chem 283(19):13205–13215

Wang Y, Toh HC, Chow P, Chung AY, Meyers DJ, Cole PA, Ooi LL, Lee CG (2012) MicroRNA-224 is up-regulated in hepatocellular carcinoma through epigenetic mechanisms. FASEB J 26(7):3032–3041

Wong QW, Lung RW, Law PT, Lai PB, Chan KY, To KF, Wong N (2008) MicroRNA-223 is commonly repressed in hepatocellular carcinoma and potentiates expression of Stathmin1. Gastroenterology 135(1):257–269

Xu T, Zhu Y, Wei QK, Yuan Y, Zhou F, Ge YY, Yang JR, Su H, Zhuang SM (2008) A functional polymorphism in the miR-146a gene is associated with the risk for hepatocellular carcinoma. Carcinogenesis 29(11):2126–2131

Xu Y, Liu L, Liu J, Zhang Y, Zhu J, Chen J, Liu S, Liu Z, Shi H, Shen H, Hu Z (2011) A potentially functional polymorphism in the promoter region of miR-34b/c is associated with an increased risk for primary hepatocellular carcinoma. Int J Cancer 128(2):412–417

Yang J, Zhou F, Xu T, Deng H, Ge YY, Zhang C, Li J, Zhuang SM (2008) Analysis of sequence variations in 59 microRNAs in hepatocellular carcinomas. Mutat Res 638(1–2):205–209

Yang MH, Chen CL, Chau GY, Chiou SH, Su CW, Chou TY, Peng WL, Wu JC (2009) Comprehensive analysis of the independent effect of Twist and Snail in promoting metastasis of hepatocellular carcinoma. Hepatology 50(5):1464–1474

Yasuda E, Kumada T, Takai S, Ishisaki A, Noda T, Matsushima-Nishiwaki R, Yoshimi N, Kato K, Toyoda H, Kaneoka Y, Yamaguchi A, Kozawa O (2005) Attenuated phosphorylation of heat shock protein 27 correlates with tumor progression in patients with hepatocellular carcinoma. Biochem Biophys Res Commun 337(1):337–342

Yoshida T, Hisamoto T, Akiba J, Koga H, Nakamura K, Tokunaga Y, Hanada S, Kumemura H, Maeyama M, Harada M, Ogata H, Yano H, Kojiro M, Ueno T, Yoshimura A, Sata M (2006) Spreds, inhibitors of the Ras/ERK signal transduction, are dysregulated in human hepatocellular carcinoma and linked to the malignant phenotype of tumors. Oncogene 25(45):6056–6066

Yoshiji H, Noguchi R, Kuriyama S, Yoshii J, Ikenaka Y, Yanase K, Namisaki T, Kitade M, Yamazaki M, Uemura M, Fukui H (2005) Different cascades in the signaling pathway of two vascular endothelial growth factor (VEGF) receptors for the VEGF-mediated murine hepatocellular carcinoma development. Oncol Rep 13(5):853–857

Zhao LJ, Wang L, Ren H, Cao J, Li L, Ke JS, Qi ZT (2005) Hepatitis C virus E2 protein promotes human hepatoma cell proliferation through the MAPK/ERK signaling pathway via cellular receptors. Exp Cell Res 305(1):23–32

Chapter 6
Animal Models of Liver Cancer

Crucial to our efforts to improve the treatment of HCC is the use of animal models that closely mimic the human condition in terms of physiology, etiology, and clinical setting. Because HCC is genetically heterogeneous, and while this is a difficult task, a number of animal models have been developed over the years for studying HCC. Studies conducted with these models are discussed herein as well as a discussion of how these models might be improved. In particular, larger animal models that might lend themselves to minimally invasive image-guided interventions on or close to the human scale are described.

Mouse

The mouse has been and remains the most preferred animal model system for cancer studies. This is due to the small size, lower cost when compared to larger animals for the same experiment, and the foundation of existing models upon which one can build. They have proved to be an invaluable tool in understanding molecular mechanisms and signal transduction pathways from a vast array of transgenic mice with mutations and knockouts of genes involved in key cellular processes such as apoptosis, cell proliferation, and differentiation (Beer et al. 2004). With respect to HCC, immune-suppressed strains with xenografts and drug-induced models are quite common. Heat transfer studies for either hyperthermia or cryoablation are extremely difficult to study on the rodent scale and visualization for percutaneous application with conventional techniques such as ultrasound is somewhat unrealistic.

© The Author(s) 2014
R.N. Aravalli, C.J. Steer, *Hepatocellular Carcinoma*, SpringerBriefs
in Cancer Research, DOI 10.1007/978-3-319-09414-4_6

Rat

It has been well established that rats continuously fed with a diet deficient in choline and methionine develop HCC in the absence of any exogenous carcinogen. One of the most common rat models of HCC is based on a choline-deficient diet. Other popular rat models of HCC involve chemical treatment to induce carcinogenesis (Lim 2002). Implantation of carcinogens rather than dietary administration has met with very limited success (Aterman 1987). While methods are available for cannulating the hepatic artery of a rat for chemoembolization for example (Li et al. 2002), these are not commonplace and the percutaneous ablation methods are only minimally improved by going from the mouse to rat.

Rabbit

There are no known models of HCC in the rabbit. Nevertheless, the VX2 tumor, based upon the work of Rous in the 1930s, has been used quite extensively for studying imaging and treatment methods in interventional oncology. It has also been used for a large variety of other tumors outside the scope of this discussion. The tumor itself has been poorly characterized in part, due to the fact that it does not grow in culture. Molecular biologists have shown little interest in the model and those using it in a variety of medical specialties may be either unfamiliar or unaffected by the limited details of its biology. Minimal histochemistry has been done, and this is mainly in support of an underlying hypothesis rather than as a freestanding investigation. The VX2 was derived from epithelial papillomas found in wild cottontail rabbits from Iowa and Kansas that were repeatedly treated with carcinogenic tars until a very aggressive anaplastic 'carcinosarcoma' developed (Rous and Beard 1934). For use, the tumor must either be propagated from live animal to live animal or with diminished success from frozen specimens. It grows quickly, rapidly develops a necrotic core, and kills the host within weeks. Its relevance to HCC is thus questionable and conclusions from studies using it must be viewed with caution.

Pig

It is well established that pigs represent a fairly close model to humans in terms of liver anatomy and physiology. They are also reasonably close in size and therefore lend themselves to human scale devices. Two relatively recent reports stand in contrast to each other in terms of their designs. The first is a drug-induced model, in which intraperitoneal dosing of diethylnitrosamine every week for a 3-month period was followed by development of cirrhosis and tumor over the course of 12–15 months (Li et al. 2006). Similar to the woodchuck model this time frame is somewhat impractical and mini-pigs would be needed to keep the size manageable. Furthermore, the disease in this model is characterized by numerous small tumors, which is uncommon in humans except for rare cases of diffuse disease presenting at a terminal state. It may be that a drug promoter such as phenobarbital would shorten the time to develop tumors, but that still leaves unanswered the issue of the extreme multiplicity inherent in this method. Based on high genetic homology between swine and human genomes, a genetically engineered pig model of tumorigenesis was produced by expressing proteins known to cause human liver cancer (Adam et al. 2007). This model produced a defined tumor in the isogenic host animal, where porcine fibroblasts expressing human proteins were transplanted in pigs after mild immune suppression. While this pig model of hepatocarcinogenesis is promising, its use in applications relevant to the human has not yet been demonstrated. Nevertheless, a pig model of liver cancer has a number of advantages that makes it ideal for preclinical studies of imaging, hyperthermia, radiation or photodynamic therapy of tumors, and needs to be explored (Adam et al. 2007).

Primate

Macaque and marmoset have both been shown to develop cancer in the liver (Foster 2005), but the relatively small size of the animals and administrative barriers to use essentially exclude these primates from any serious considerations. Furthermore, especially with drug-induced neoplasia, there is mounting evidence from metabolic and toxicologic studies that non-human primates do not necessarily mimic human disease.

Woodchuck

Hepatomas from Woodchuck Hepatitis Virus (WHV) were first reported in the mid 1950s with more work reported in the late 1960s and early 1970s. The disease is considered closely related to hepatitis B in humans. Currently there is a colony at Cornell that reliably develops tumor over the course of 2–4 years provided there is a large enough viral load in the inoculum of WHV at birth (Tennant et al. 2004; Kulkarni et al. 2007). Given the time course for disease development this may be useful for mechanistic and preventive studies but from a treatment and outcome standpoint it is impractical. Cirrhosis does not typically develop in these animals and, therefore it does not represent an entirely accurate animal model of a human disease.

A Case for Better Animal Models

In order to reduce morbidity and mortality from HCC, improvements in early diagnosis and development of novel local and systemic therapies for advanced disease are essential. Animal models are critical tools not only to achieve these objectives but also to decipher the genotype-phenotype relationship based upon risk factors and host's genetic and hereditary traits. Although they have been useful in elucidating certain molecular mechanisms of HCC, there are no animal models that recapitulate the human condition. Therefore, there is a pressing need to focus our intention on generating such a model that should (1) allow accurate assessment of treatment effects; (2) be reliable and predictable in tumor generation and studies on growth kinetics; (3) be easily imaged by conventional radiological methods; (4) be inexpensive and avoid any requirement for immune suppression; (5) allow for survival differences to manifest (no short term mortality from spontaneous or iatrogenic metastatic disease); and (6) require the tumor to be derived from a relevant cell type that lends itself to easy propagation and characterization in cell culture. Although demanding, a model that meets these criteria would allow us to study the human disease more accurately and to develop novel therapeutics that can be directly translated to the bedside (Aravalli et al. 2009).

References

Adam S, Rund LA, Kuzmuk KN, Zachary JF, Schook LB, Counter CM (2007) Genetic induction of tumorigenesis in swine. Oncogene 26:1038–1045

Aravalli RN, Golzarian J, Cressman EN (2009) Animal models of cancer in interventional radiology. Eur Radiol 19(5):1049–1053

Aterman K (1987) Localized hepatocarcinogenesis: the response of the liver and kidney to implanted carcinogens. J Cancer Res Clin Oncol 1:507–538

Beer S, Zetterberg A, Ihrie RA, McTaggart RA, Yang Q, Bradon N, Arvanitis C, Attardi LD, Feng S, Ruebner B, Cardiff RD, Felsher DW (2004) Developmental context determines latency of MYC-Induced tumorigenesis. PLoS Biol 2(11):1785–1798

Foster J (2005) Spontaneous and drug-induced hepatic pathology of the laboratory beagle dog, the cynomolgus macaque and the marmoset. Toxicol Pathol 33(1):63–74

Kulkarni K, Jacobson IM, Tennant BC (2007) The role of the Woodchuck model in the treatment of hepatitis B virus infection. Clin Liver Dis 11:707–725

Li X, Zheng C-S, Feng G-S, Zhao C-K, Zhao J-G, Liu X (2002) An implantable rat liver tumor model for experimental transarterial chemoembolization therapy and its imaging features. World J Gastroenterol 8(6):1035–1039

Li X, Zhou X, Guan Y, Wang Y-X J, Scutt D, Gong Q-Y (2006) N-Nitrosodiethylamine-induced pig liver hepatocellular carcinoma model: radiological and histopathological studies. Cardiovasc Intervent Radiol 29:420–428

Lim IK (2002) Spectrum of molecular changes during hepatocarcinogenesis induced by DEN and other chemicals in Fischer 344 male rats. Mech Ageing Dev 123(12):1665–1680

Rous P, Beard JW (1934) A virus-induced mammalian growth with the characters of a tumor (the Shope Rabbit Papilloma): II Experimental alterations of the growth on the skin: morphological considerations: the phenomena of retrogression. J Exp Med 60(6):723–740

Tennant B, Toshkov IA, Peek SF, Jacob JR, Menne S, Hornbuckle WE, Schinazi RD, Korba BE, Cote PJ, Gerin JL (2004) Hepatocellular carcinoma in the Woodchuck model of hepatitis B virus infection. Gastroenterology 127:S283–S293

Chapter 7
Novel Therapeutic Strategies to Combat HCC

Current treatment options for advanced HCC are limited. HCC is resistant to conventional chemotherapy, with fairly common response rates <20 %, no overall survival benefits, and no cures. In recent years, however, remarkable progress has been made in understanding the molecular basis of hepatocarcinogenesis. While most of these approaches have relied on interfering with different signaling pathways, others have focused on modulating various cellular processes (Tables 7.1 to 7.3) (Pang and Poon 2007; Llovet and Bruix 2008). These efforts have resulted in the identification of novel molecular biomarkers and the design of therapeutic drugs. Among these, Sorafenib (previously known as BAY 43-9006) (Wilhelm et al. 2006) has significantly changed the landscape in the clinical management of HCC. Indeed, after the dismal results of prior efforts it could be considered a turning point, even if the benefits are rather modest and potential side effects limit its use. Adding to the difficulty is that it is impossible to predict at this point which patients will benefit from the drug (Aravalli and Cressman 2009).

Small Molecule-Based Therapeutics

To date, Sorafenib [N-(3-trifluoromethyl-4-chlorophenyl)-N'-(4-(2-methylcarbamoyl pyridin-4-yl)oxyphenyl)urea] is the only drug approved for the treatment of HCC in humans by the Food and Drug Administration in the United States. It was originally identified as an inhibitor of Raf-1, a member of the RAF/MEK/ERK signaling pathway, to suppress tumor cell line proliferation, and growth in several xenograft models (Lyons et al. 2001; Wilhelm et al. 2004). In later studies, it was characterized as a multikinase inhibitor that blocks proliferation and angiogenesis of tumors not only by suppressing the Raf/MEK/ERK signaling pathway, but also by inhibiting the activities of various receptor tyrosine kinases of growth factor receptors, including VEGFR2, PDGFR, FLT3, Ret, and c-Kit (Liu et al. 2006).

© The Author(s) 2014
R.N. Aravalli, C.J. Steer, *Hepatocellular Carcinoma*, SpringerBriefs
in Cancer Research, DOI 10.1007/978-3-319-09414-4_7

Many studies have investigated the biological functions of Sorafenib in HCC cell lines, tumor xenografts, and in animal models of HCC. Sorafenib treatment of rat and human HCC cell lines decreased the phosphorylation of STAT3 at tyrosine (Y705) and serine residues (S727), but it did not affect the phosphorylation of Janus kinase 2 (JAK2) and shatterproof 2 (SHP2) phosphatase (Gu et al. 2011). Dephosphorylation of Y705 and S727 were associated with reduced phosphorylation of Akt and ERK, respectively. Sorafenib treatment led to accumulation of autophagosomes and inhibition of mammalian target of rapamycin (mTOR) complex 1 in HuH7, HLF and PLC/PRF/5 HCC cell lines (Shimizu et al. 2012) In PLC/PRF/5 and HepG2 cells, it inhibited the phosphorylation of MEK and ERK and down-regulated cyclin D1 levels (Liu et al. 2006) Sorafenib treatment induced differential expression of 19 proteins in the proteome of HepG2 cells (Suo et al. 2012). Among these, annexin A1 and cyclophilin A were significantly downregulated, suggesting their potential oncogenic role in HCC tissues. Even though Sorafenib did prolong the survival time in HCC patients, some adverse events have been reported in patients receiving this drug. These include hand-foot skin reactions (HFSR), and hyperbilirubinaemia associated with marked elevations of alanine aminotransferase (ALT) (Chen et al. 2010). These adverse reactions have either resulted in dosage reduction or in the administration of Sorafenib with other drugs.

The B cell lymphoma extra-large (bcl-xl) inhibitor ABT737 induced apoptosis, and suppressed growth of HuH7 xenograft tumors in mice when combined with Sorafenib (Hikita et al. 2010). In a recent study, the combination of K vitamins with Sorafenib induced c-Raf phosphorylation at two serine residues, Ser-43 and Ser-259 and this significantly inhibited HCC growth (Carr et al. 2011). The authors found that vitamin K1 enhanced sorafenib-induced c-Met phosphorylation at Tyr-1349, and consequently induced phosphorylation of PI3K-Akt. These findings suggested that the c-Met-PI3K-Akt signaling pathway to inhibit c-Raf phosphorylation may play a central role in inhibiting HCC cell growth when vitamin K1 is used in combination with Sorafenib (Carr et al. 2011).

The success of Sorafenib has resulted in the development of a number of therapeutic compounds, some of which are currently undergoing clinical trials (Tables 2-4). While these drugs have been shown to inhibit their target pathway in experimental models, others were tested in cell lines and preclinical models. Compounds that target other cellular proteins are currently under development (Aravalli et al. 2013). Integration of molecular biomarkers identified from various different studies with these drug targets would allow us to identify potential candidates for drug response.

Several studies have shown that heparin-degrading endosulfatases, SULF1 and SULF2, play important roles in modulating heparin-binding growth signaling pathways (Leonardi et al. 2012). Even though these proteins are structurally similar, they have opposing effects on FGF signaling and its downstream Akt/MAPK pathway. SULF-2-dependent desulfation of HSPGs releases growth factors from storage, and increases binding to their receptors, leading to the activation of growth signaling (Lai et al. 2008; Leonardi et al. 2012). A heparan sulfate mimetic, PI-88, that has been shown to inhibit the activity of SULFs is currently undergoing phase

Table 7.1 Molecular targeted therapies currently under clinical trials

Target signaling pathway	Drug/compound[a]
VEGFR	Sorafenib, Bevacizumab, Brivanib, Ramucirumab, Sunitinib, Vatalanib, Foretinib, Regorafenib, SU5416, TSU-68, E7080, Dovitinib, Vandetanib, Cediranib, Pazopanib, BIBF1120, IMC-1121B, XL84, Linifanib
PDGFR	Sorafenib, Bevacizumab, Brivanib, Sunitinib, Vatalanib, Regorafenib, SU5416, TSU-68, Dovitinib, Cediranib, Pazopanib, BIBF1120, Imatinib, Linifanib
IGFR	BIIB022, XL-228, PPP, AVE1642, IMC-A12, Cixutumumab, OSI-906, OSI-906, AMG386
FGFR	Brivanib, Sunitinib, TSU-68, Dovitinib, BIBF1120, PI-88
EGFR	Vandetanib, Erlotinib, Gefitinib, Cetuximab, Lapanitib

[a]These compounds include small molecules, antisense oligonucleotides, and monoclonal antibodies

II clinical trial, as an adjuvant therapy for HCC after curative resection (Liu et al. 2009; Hossain et al. 2010). A number of antibodies and other drugs that target various growth factor signaling pathways are also currently being evaluated for their safety and efficacy in various clinical trials (Tables 7.1, 7.2, 7.3).

Immunotherapy is an emerging area of research for HCC. Although immune-based therapeutics are potentially effective, distinct immunosuppressive milieu of both the steady-state and diseased liver presented a major obstacle for these studies (Pardee and Butterfield 2012). In addition, significant variation in the heterogeneity of the molecular and cellular makeup of HCC due to multiple etiological factors has posed significant problems. Nevertheless, efforts are underway to develop locally applied oncolytic viral agents as well as more DNA, peptide, viral, and dendritic cell-based vaccines for HCC (Rinaldi et al. 2009; Zhang et al. 2009; Pardee and Butterfield 2012). In a recent study, DC-mediated CTLs were effectively targeted to CSCs in HCC (Sun et al. 2010).

Over the past decade, remarkable progress has been made in terms of our understanding of the molecular basis of liver cancer. This advancement has resulted in the identification of novel molecular biomarkers and the design of therapeutic drugs. The most intense investigations have centered on signal transduction, apoptosis, and angiogenesis. While most of these approaches have relied on interfering with various signaling pathways, others have focused on blocking neoangiogenesis (Llovet and Bruix 2008; Pang and Poon 2007). The now FDA-approved compound Sorafenib has been shown to block angiogenesis as a key mechanism of inhibiting the growth of HCC (Wilhelm et al. 2006). Indeed, after the abysmal results of prior efforts it could be considered a turning point even if the benefits are rather modest and side effects limit its use in many instances. Unfortunately, it is impossible to predict at this point which patients will benefit from the drug.

Sorafenib is multikinase inhibitor that was shown to inhibit tumor angiogenesis by targeting the Raf/MERK/ERK pathway in renal cell carcinoma (Adnane et al. 2006). It was later shown not only to target this pathway but also to induce apoptosis

Table 7.2 Clinical trials targeting specific cellular targets

Target	Drug/compound[a]
HDAC	Panobinostat, Vorinostat, LBH589, Belinostat, Resminostat
mTOR	Sirolimus, AZD8055, Temsirolimus, Everolimus, Rapamycin
c-kit	Sorafenib, Regorafenib, Pazopanib
c-met	Foretinib, XL84
TIE-2	Regorafenib
Dipeptidyl peptidases	Talabostat
MEK	AZD6344, AZD6244
TRAIL	Mapatumumab, CS-1008
XIAP	AEG35156
Proteasome	Bortezomib
CDK	Alvocidib
PD1	CT-011
Survivin	LY2181308
RAR	Z-208, TAC-101
PI3K/Akt	Eveolimus, MK-2206, PI-88
HER2/neu	Lapatinib
BCR-ABL	Dasatinib
Farnesyl-OH-transferase	Lonafarnib
Glypican-3	GC33
Angiopoietin	AMG386
HAb18G/CD147	Licartin
Caspases	IDN-6558
Aurora kinase	MLN8237
Kinesin spindle protein	Ispinesib
Bcl-2	Oblimersen
CSF-1R	Linifanib, Sunitinib
MAPK	Selumetinib

[a]These compounds include small molecules, antisense oligonucleotides, and monoclonal antibodies

in the PLC/PRF/5 xenograft model of HCC (Liu et al. 2006). The success of Sorafenib has resulted in the development of a number of therapeutic compounds, some of which are undergoing clinical trials (Tables 7.1–7.3). These drugs have been shown to inhibit their target pathway in cell lines, experimental and preclinical models. Compounds that target the Wnt signaling pathway are currently under development. Integration of molecular biomarkers identified from various different studies with these drug targets would allow us to identify potential candidates for drug response. A final area of therapy is immunotherapy; and there are ongoing efforts to develop locally applied oncolytic viral agents for tumor destruction as well as more conventional efforts towards vaccination against tumors (Rinaldi et al. 2009; Tian et al. 2009; Zhang et al. 2009).

Table 7.3 Combination therapies currently under clinical trials

Drug combination[a]	Target pathway/molecule
Sorafenib + Panobinostat	VEGFR, PDGFR, BRAF, c-kit, Raf, HDAC
Sorafenib + BIIB022	VEGFR, PDGFR, BRAF, c-kit, Raf, IGF-1R
Sorafenib + Cixutumumab	VEGFR, PDGFR, BRAF, c-kit, Raf, IGF-1R
Sorafenib + ARQ197	VEGFR, PDGFR, BRAF, c-kit, Raf, c-Met
Sorafenib + BIBF1120	VEGFR, PDGFR, BRAF, c-kit, Raf, FGFR
Sorafenib + S-1	VEGFR, PDGFR, BRAF, c-kit, Raf
Sorafenib + PR-104	VEGFR, PDGFR, BRAF, c-kit, Raf
Sorafenib + Temsirolimus	VEGFR, PDGFR, BRAF, c-kit, Raf, mTOR
Sorafenib + AZD6244	VEGFR, PDGFR, BRAF, c-kit, Raf, MEK
Sorafenib + AVE1642	VEGFR, PDGFR, BRAF, c-kit, Raf, IGF-1R
Erlotinib + Celecoxib	EGFR
Erlotinib + AVE1642	EGFR, IGF-1R
Erlotinib + Bevacizumab	EGFR, VEGFR
Erlotinib + Docetaxel	EGFR
Erlotinib + GEMOX	EGFR
Bevacizumab + GEMOX	EGFR, VEGFR
Bevacizumab + TACE	VEGFR
Bevacizumab + CAPEOX	VEGFR
Cetuximab + CAPEOX	EGFR
Gemcitabine + Docetaxel	ATM, ATR, Chk1, Chk2
Sorafenib + TACE	VEGFR, PDGFR, BRAF, c-kit, Raf
Sorafenib + Erlotinib	VEGFR, PDGFR, BRAF, c-kit, Raf, EGFR
Sorafenib + OSI-906	VEGFR, PDGFR, BRAF, c-kit, Raf, IGF-1R, IR

Modified from Aravalli et al. (2013), with permission, Springer
[a]These compounds include small molecules, antisense oligonucleotides, and monoclonal ntibodies
Abbreviations: *GEMOX* Gemcitabine-Oxaliplatin, *CAPEOX* Capecitabine-Oxaliplatin, *TACE* Transarterial chemoembolization

MicroRNA Therapeutics

The field of miRNA therapeutics for HCC is still in its infancy; and to date, only a handful of successful outcomes have been reported. By using the above described approaches that were successful with other cancers, experiments can be designed to suppress oncomiR expression, and to restore the normal expression of dysregulated miRNAs in tumor tissues of the liver. Some of the potential candidates for these miRNA therapeutics are discussed below.

Inhibition of OncomiRs

Oncomirs promote tumor growth and proliferation by blocking the activities of cellular tumor suppressors, cell cycle regulators and pro-apoptotic genes. For instance, miR-224 increases tumor cell progression by inhibiting the expression of apoptosis inhibitor-5 (Wang et al. 2008, 2012a); and miR-221 promotes cell proliferation by controlling the cell cycle inhibitors CDKN1C/p57 and CDKN1B/p27 (Fornari et al. 2008). miR-519d, on the other hand, contributes to HCC by targeting CDKNA/p21, PTEN, Akt3 and TIMP2 (Fornari et al. 2012). Intratumoral administration of miR-143 showed that high levels of miR-143 can significantly promote tumor metastasis in an athymic nude mouse model by repressing the expression of fibronectin type III domain containing 3B (FNDC3B), which regulates cell motility (Zhang et al. 2009). miR-222, a frequently overexpressed miRNA in HCC, also regulates cell motility by enhancing Akt signaling (Wong et al. 2010), whereas miR-423 directly binds to 3′UTR of p21Cip1/Waf1 to suppress its expression and to promote cell cycle progression, and tumorigenesis (Lin et al. 2011).

While miR-146a increases the resistance of HCC cells from cytotoxic effects of interferon-α by inhibiting the expression of Smad4 (Tomokuni et al. 2011), upregulation of hepatic transforming growth factor (TGF)-β and its downstream mediators Smads 2, 3 and 4, was found to be correlated with an increased expression of miR-181 in the livers of mice fed with choline-deficient L-amino acid-defined (CDAA) diet (Wang et al. 2010). Depletion of miR-181b inhibited tumor growth in nude mice, whereas its expression enhanced resistance of HCC cells to the anticancer drug doxorubicin (Wang et al. 2010). miR-17-5p up-regulates the migration and proliferation of HCC cells by activating the p38 mitogen-activated protein kinase MAPK pathway and increasing the phosphorylation of heat shock protein 27 (Yang et al. 2010), suggesting its potential application in HCC therapy. The miR-34 family members are direct transcriptional targets of tumor suppressor p53, and loss of miR-34 function can impair p53-mediated cell cycle arrest and apoptosis (Xu et al. 2011a).

miR-210, which is often upregulated in HCC, promotes hypoxia-mediated tumor cell metastasis by targeting the vacuole membrane protein 1 (Ying et al. 2011). Therefore, inhibition of miR-210 expression could lead to reduction in HCC metastasis, and should facilitate the development of novel therapeutic strategy against hypoxic tumor cells. Similarly, miR-30d, a miRNA associated with intrahepatic metastasis of HCC, promotes tumor cell migration and invasion *in vitro* and intrahepatic and distal pulmonary metastasis *in vivo* by targeting the Galphai2 (GNAI2) protein (Yao et al. 2010). The expression of miR-148a is elevated in HepG2 and Hep3B cells where it promotes cell proliferation, cell cycle progression, and cell migration. When anti-miR-148a was overexpressed from a lentivirus, it inhibited the activity of miR-148a, in addition to blocking the Akt signaling pathway in these cells, suggesting that it could serve as an early diagnostic marker and/or therapeutic target (Yuan et al. 2012). Additionally, miR-1 and miR-499 inhibited invasion and migration of HepG2 cells by targeting the ets1 proto-oncogene, which causes degradation of extracellular matrix (Wei et al. 2012). Another antagomir iR-219-5p induced tumor suppressive effects in HCC cell lines by targeting glypican-3 and by causing cell cycle arrest at the G1 to S transition (Huang et al. 2012).

While successful strategies to inhibit oncomirs for HCC therapy have not yet been reported, each of these miRNAs represent a strong candidate for therapeutic intervention towards achieving a gain of function with the use of miRNA antagonists, RNAi, and small molecule inhibitors. Further in-depth studies on these candidate miRNAs will allow us to ascertain their potential as targets for the treatment of HCC.

MicroRNA Replacement Therapy

The first successful demonstration of miRNA replacement to restore the expression levels of a downregulated miRNA was reported using a miR-26a-encoding AAV vector in a mouse model of HCC (Kota et al. 2009). miR-26a is normally decreased in HCC, and its overexpression in mouse livers resulted in the inhibition of cancer cell proliferation, and induction of tumor-specific apoptosis. More recently, another miRNA downregulated in HCC, miR-34a, has been successfully tested using this approach and delivering it with NOV340 liposomes in an orthotopic model of HCC (Bader 2012). In both cases, significant tumor reduction, dramatic protection from disease progression without toxicity, and prolonged survival of animals was reported. The combination of both restoring and inhibiting specific miRNAs will be critical in developing effective therapeutic agents for HCC. Each of these studies was based on the regulation of multiple cellular pathways associated with human disease, which now appears to be a requirement for successful cancer therapy (Bader 2012). Along these lines, another downregulated miRNA in HCC, miR-375 was overexpressed in liver cancer cells. It decreased cell proliferation, clonogenicity, migration/invasion, induced G1 arrest and apoptosis (He et al. 2012) *in vitro* indicating that it could be a potential candidate for *in vivo* studies. Indeed, when cholesterol-conjugated $2'$-O-methyl-modified miR-375 mimics (Chol-miR-375) were administered in nude mice, they were able to significantly suppress the growth of hepatoma xenografts (He et al. 2012).

In other promising studies, overexpression of miR-376a suppressed cell proliferation in HuH7 cells (Zheng et al. 2012), and miR-138 induced cell cycle arrest by targeting cyclin D3 (Wang et al. 2012b). In an *in vitro* experiment with HCC cell lines transfected with miR-637 mimics or transduced with Lv-miR637 vectors, miR-637 suppressed tumor growth by negatively regulating the phosphorylation of STAT3 protein (Zhang et al. 2011). Downregulation of miR-29b correlated with rapid recurrence and poor survival of individuals with HCC. miR-29b dramatically suppressed the ability of HCC cells to promote capillary tube formation of endothelial cells and to invade extracellular matrix gel *in vitro*; and in mouse xenografts where it regulated the expression of matrix metalloproteinase 2 (Fang et al. 2011). Restoration of miR-29b expression resulted in significantly reduced angiogenesis, invasion, and metastasis of HCC (Fang et al. 2011), suggesting its potential use as a novel therapeutic target in anti-HCC therapy.

Intratumoral injection of cholesterol-conjugated miR-99a mimics was recently demonstrated to significantly inhibit tumor growth and to reduce α-fetoprotein levels in HCC-bearing nude mice. In this study, miR-99a expression inversely correlated with protein levels of insulin-like growth factor 1 receptor (IGF-1R) and mammalian target of rapamycin (mTOR), inducing cell cycle arrest at the G1 phase (Li et al. 2011). This finding suggested a potential tumor suppressor role for miR-99a. Another miRNA often downregulated in HCC is miR-199a-3p. By transfecting pre-miR-199a-3p and anti-miR-199a-3p oligonucleotides, it was shown to target mTOR and c-met in HCC cells (Fornari et al. 2010). Restoring attenuated levels of miR-199a-3p in these cells led to cell cycle arrest at the G1 phase, reduced invasive capability, enhanced susceptibility to hypoxia, and increased sensitivity to doxorubicin-induced apoptosis, suggesting that enhancement of miR-199a-3p levels may have therapeutic benefits in HCC (Fornari et al. 2010).

miR-122 binds to the 3′-UTR of Bcl-2 family member Bcl-w, an anti-apoptotic protein, to induce cell death in HepG2 and Hep3B cells (Lin et al. 2008). When expressed from an Ad vector, miR-122 sensitized HCC cells to adriamycin and vincristine treatment by causing cell cycle arrest at the G2/M phase and by modulating the expression of multidrug resistant protein MDR-1 (Xu et al. 2011b). Using several 3′ cholesterol-conjugated, 2′-O-Me oligonucleotides, and an unconjugated 2′-O-methoxyethyl–phosphorothioate-modified oligonucleotide, miR-122 was inhibited in mouse livers to demonstrate its function in lipid metabolism (Krützfeldt et al. 2005; Esau et al. 2006). The systemic administration of a 16-mer LNA-modified anti-miR oligonucleotide was also reported to significantly inhibit miR-122 in mouse liver (Elmén et al. 2008). These studies provide strong evidence that miR-122-based HCV therapy can be developed with antagomirs. A LNA-modified phosphorothioate oligonucleotide (SPC3649) complementary to the 5′-end of miR-122 was administered to chimpanzees with chronic HCV infection, and showed that SPC3649 suppression of miR-122 had long-lasting suppression of viraemia with no evidence for viral resistance or side effects in the treated animals (Lanford et al. 2010). In addition to demonstrating the feasibility and safety of prolonged administration of a LNA oligonucleotide drug *in vivo*, this study showed that miR-122 is essential for HCV accumulation, and that the miR-122 seed sites were conserved in all HCV genotypes and subtypes suggesting genotype independent nature of this therapy. This was the first miRNA-based therapeutic drug developed to treat a liver disease, and was tested by Santaris Pharma (Hørsholm, Denmark) in phase 2 clinical trials (Lindow and Kauppinen 2012). More recently, it was reported that HCV patients were successfully treated with an oligonucleotide drug MiraVirsen that targets miR-122 (Janssen et al. 2013).

These are some of the promising candidates for miRNA replacement therapy for HCC. It is important to exercise caution while designing clinical trials with miRNA mimics for targeting multiple genes relevant to human disease. There are always concerns about potential toxicity in normal tissues, especially under conditions where the therapeutic delivery of miRNA mimics will also lead to an accumulation of exogenous miRNA in normal cells. These toxic effects might be the result of overloading RISC with the exogenous miRNA, thereby competing with endogenous miRNAs necessary for normal cellular welfare, and/or hyperactivating cellular pathways that will also reduce the viability of normal cells (Bader et al. 2010).

Stem Cell Therapy

Perhaps the most important implication of HCC stem cells is their potential clinical impact in developing novel therapeutic approaches for HCC. Recently, several groups have reported isolation and characterization of human HCC stem cells. For example, CD133 which has been reported to be a marker of cancer stem cells in various tissues (brain, pancreas, prostate, colon) was also used to identify cancer stem cells in hepatocellular carcinoma cell lines (Ma et al. 2007). They found that HCC was hierarchically organized and originated from a population of progenitor cells that expressed CD133$^+$. These cells isolated from human HCC cell lines represented only a minority of the tumor cell population in human HCC specimens, and they had a greater colony-forming efficiency, higher proliferation potential and greater ability to form tumor in animal models. They also possessed characteristics similar to those of stem cells including a gene expression profile, the ability to self-renew and the ability to differentiate. In a follow-up paper, the same group showed that CD133$^+$ HCC stem cells were the cell population in HCC responsible from chemotherapy resistance (to doxorubicin and 5-fluorouracil) and could be the source of tumor recurrence after chemotherapy (Ma et al. 2008). They demonstrated that CD133$^+$ HCC cells survived chemotherapy significantly better than most tumor cells that did not express CD133; and the underlying mechanism was the constitutive activation of the Akt/PKB and Bcl-2 cell survival pathways. An obvious clinical implication of this finding is the fact that specific inhibitors of these pathways would potentially be useful in the treatment of HCC.

In another study, CD90 expression was used as a marker to characterize HCC stem cells (Yang et al. 2008). In this study, HCC stem cells were isolated from hepatocellular carcinoma cell lines, tumor specimens, and blood samples. The investigators showed that CD45$^-$ and CD90$^+$ cells, but not the CD90$^-$ cells, from HCC cell lines had tumorigenic potential. All of the tumor specimens and virtually all of the blood samples from patients with HCC contained a population of CD90$^+$ and CD44$^-$ cells that were capable of generating tumor nodules in immunodeficient mice. The CD90$^+$CD44$^+$ cells demonstrated a more aggressive phenotype than the CD90$^+$CD44$^-$ cells and formed metastatic lesions in the lungs of immunodeficient mice. The gene expression profile of these CD45$^-$CD90$^+$ cells indicated that they also had stem cell-like phenotype. Blocking CD44 expression prevented the formation of local and metastatic tumor nodules by the CD90$^+$ cells suggesting that targeting CD44 might be useful in the treatment of HCC.

These studies illustrate the importance of identifying and characterizing HCC stem cells. The information gained through such studies will not only further our understanding of the development and progression of hepatic carcinoma, but also will pave the way for the development of more specific and more effective treatment strategies for HCC.

References

Adnane L, Trail PA, Taylor I, Wilhelm SM (2006) Sorafenib (BAY 43-9006, Nexavar), a dual-action inhibitor that targets RAF/MEK/ERK pathway in tumor cells and tyrosine kinases VEGFR/PDGFR in tumor vasculature. Methods Enzymol 407:597–612

Aravalli RN, Cressman EN (2009) Molecular signaling in hepatocelluar carcinoma. Cancer Chemother Rev 4:157–164

Aravalli RN, Cressman EN, Steer CJ (2013) Cellular and molecular mechanisms of hepatocellular carcinoma: an update. Arch Toxicol 87(2):227–247

Bader AG (2012) miR-34 – a microRNA replacement therapy is headed to the clinic. Front Genet 3:120

Bader AG, Brown D, Winkler M (2010) The promise of microRNA replacement therapy. Cancer Res 70(18):7027–7030

Carr BI, Wang Z, Wang M, Cavallini A, D'Alessandro R, Refolo MG (2011) c-Met-Akt pathway-mediated enhancement of inhibitory c-Raf phosphorylation is involved in vitamin K1 and sorafenib synergy on HCC growth inhibition. Cancer Biol Ther 12(6):531–538

Chen PJ, Furuse J, Han KH, Hsu C, Lim HY, Moon H, Qin S, Ye SL, Yeoh EM, Yeo W (2010) Issues and controversies of hepatocellular carcinoma-targeted therapy clinical trials in Asia: experts' opinion. Liver Int 30(10):1427–1438

Elmén J, Lindow M, Silahtaroglu A, Bak M, Christensen M, Lind-Thomsen A, Hedtjärn M, Hansen JB, Hansen HF, Straarup EM, McCullagh K, Kearney P, Kauppinen S (2008) Antagonism of microRNA-122 in mice by systemically administered LNA-antimiR leads to up-regulation of a large set of predicted target mRNAs in the liver. Nucleic Acids Res 36(4):1153–1162

Esau C, Davis S, Murray SF, Yu XX, Pandey SK, Pear M, Watts L, Booten SL, Graham M, McKay R, Subramaniam A, Propp S, Lollo BA, Freier S, Bennett CF, Bhanot S, Monia BP (2006) miR-122 regulation of lipid metabolism revealed by *in vivo* antisense targeting. Cell Metab 3(2):87–98

Fang JH, Zhou HC, Zeng C, Yang J, Liu Y, Huang X, Zhang JP, Guan XY, Zhuang SM (2011) MicroRNA-29b suppresses tumor angiogenesis, invasion, and metastasis by regulating matrix metalloproteinase 2 expression. Hepatology 54(5):1729–1740

Fornari F, Gramantieri L, Ferracin M, Veronese A, Sabbioni S, Calin GA, Grazi GL, Giovannini C, Croce CM, Bolondi L, Negrini M (2008) MiR-221 controls CDKN1C/p57 and CDKN1B/p27 expression in human hepatocellular carcinoma. Oncogene 27(43):5651–5661

Fornari F, Milazzo M, Chieco P, Negrini M, Calin GA, Grazi GL, Pollutri D, Croce CM, Bolondi L, Gramantieri L (2010) MiR-199a-3p regulates mTOR and c-Met to influence the doxorubicin sensitivity of human hepatocarcinoma cells. Cancer Res 70(12):5184–5193

Fornari F, Milazzo M, Chieco P, Negrini M, Marasco E, Capranico G, Mantovani V, Marinello J, Sabbioni S, Callegari E, Cescon M, Ravaioli M, Croce CM, Bolondi L, Gramantieri L (2012) In hepatocellular carcinoma miR-519d is up-regulated by p53 and DNA hypomethylation and targets CDKN1A/p21, PTEN, AKT3 and TIMP2. J Pathol 227(3):275–285

Gu FM, Li QL, Gao Q, Jiang JH, Huang XY, Pan JF, Fan J, Zhou J (2011) Sorafenib inhibits growth and metastasis of hepatocellular carcinoma by blocking STAT3. World J Gastroenterol 17(34):3922–3932

He XX, Chang Y, Meng FY, Wang MY, Xie QH, Tang F, Li PY, Song YH, Lin JS (2012) MicroRNA-375 targets AEG-1 in hepatocellular carcinoma and suppresses liver cancer cell growth *in vitro* and *in vivo*. Oncogene 31(28):3357–3369

Hikita H, Takehara T, Shimizu S, Kodama T, Shigekawa M, Iwase K, Hosui A, Miyagi T, Tatsumi T, Ishida H, Li W, Kanto T, Hiramatsu N, Hayashi N (2010) The Bcl-xL inhibitor, ABT-737, efficiently induces apoptosis and suppresses growth of hepatoma cells in combination with Sorafenib. Hepatology 52(4):1310–1321

Hossain MM, Hosono-Fukao T, Tang R, Sugaya N, van Kuppevelt TH, Jenniskens GJ, Kimata K, Rosen SD, Uchimura K (2010) Direct detection of HSulf-1 and HSulf-2 activities on extracellular heparan sulfate and their inhibition by PI-88. Glycobiology 20(2):175–186

Huang N, Lin J, Ruan J, Su N, Qing R, Liu F, He B, Lv C, Zheng D, Luo R (2012) MiR-219-5p inhibits hepatocellular carcinoma cell proliferation by targeting glypican-3. FEBS Lett 586(6):884–891

Janssen HL, Reesink HW, Lawitz EJ, Zeuzem S, Rodriguez-Torres M, Patel K, van der Meer AJ, Patick AK, Chen A, Zhou Y, Persson R, King BD, Kauppinen S, Levin AA, Hodges MR (2013) Treatment of HCV infection by targeting microRNA. N Engl J Med 368(18):1685–1694

Kota J, Chivukula RR, O'Donnell KA, Wentzel EA, Montgomery CL, Hwang HW, Chang TC, Vivekanandan P, Torbenson M, Clark KR, Mendell JR, Mendell JT (2009) Therapeutic microRNA delivery suppresses tumorigenesis in a murine liver cancer model. Cell 137(6):1005–1017

Krützfeldt J, Rajewsky N, Braich R, Rajeev KG, Tuschl T, Manoharan M, Stoffel M (2005) Silencing of microRNAs *in vivo* with 'antagomirs'. Nature 438(7068):685–689

Lai JP, Sandhu DS, Yu C, Han T, Moser CD, Jackson KK, Guerrero RB, Aderca I, Isomoto H, Garrity-Park MM, Zou H, Shire AM, Nagorney DM, Sanderson SO, Adjei AA, Lee JS, Thorgeirsson SS, Roberts LR (2008) Sulfatase 2 up-regulates glypican 3, promotes fibroblast growth factor signaling, and decreases survival in hepatocellular carcinoma. Hepatology 47(4):1211–1222

Lanford RE, Hildebrandt-Eriksen ES, Petri A, Persson R, Lindow M, Munk ME, Kauppinen S, Ørum H (2010) Therapeutic silencing of microRNA-122 in primates with chronic hepatitis C virus infection. Science 327(5962):198–201

Leonardi GC, Candido S, Cervello M, Nicolosi D, Raiti F, Travali S, Spandidos DA, Libra M (2012) The tumor microenvironment in hepatocellular carcinoma (review). Int J Oncol 40(6):1733–1747

Li D, Liu X, Lin L, Hou J, Li N, Wang C, Wang P, Zhang Q, Zhang P, Zhou W, Wang Z, Ding G, Zhuang SM, Zheng L, Tao W, Cao X (2011) MicroRNA-99a inhibits hepatocellular carcinoma growth and correlates with prognosis of patients with hepatocellular carcinoma. J Biol Chem 286(42):36677–36685

Lin CJ, Gong HY, Tseng HC, Wang WL, Wu JL (2008) miR-122 targets an anti-apoptotic gene, Bcl-w, in human hepatocellular carcinoma cell lines. Biochem Biophys Res Commun 375(3):315–320

Lin J, Huang S, Wu S, Ding J, Zhao Y, Liang L, Tian Q, Zha R, Zhan R, He X (2011) MicroRNA-423 promotes cell growth and regulates G(1)/S transition by targeting p21Cip1/Waf1 in hepatocellular carcinoma. Carcinogenesis 32(11):1641–1647

Lindow M, Kauppinen S (2012) Discovering the first microRNA-targeted drug. J Cell Biol 199(3):407–412

Liu L, Cao Y, Chen C, Zhang X, McNabola A, Wilkie D, Wilhelm S, Lynch M, Carter C (2006) Sorafenib blocks the RAF/MEK/ERK pathway, inhibits tumor angiogenesis, and induces tumor cell apoptosis in hepatocellular carcinoma model PLC/PRF/5. Cancer Res 66(24):11851–11858

Liu CJ, Lee PH, Lin DY, Wu CC, Jeng LB, Lin PW, Mok KT, Lee WC, Yeh HZ, Ho MC, Yang SS, Lee CC, Yu MC, Hu RH, Peng CY, Lai KL, Chang SS, Chen PJ (2009) Heparanase inhibitor PI-88 as adjuvant therapy for hepatocellular carcinoma after curative resection: a randomized phase II trial for safety and optimal dosage. J Hepatol 50(5):958–968

Llovet JM, Bruix J (2008) Molecular targeted therapies in hepatocellular carcinoma. Hepatology 48(4):1312–1327

Lyons JF, Wilhelm S, Hibner B, Bollag G (2001) Discovery of a novel Raf kinase inhibitor. Endocr Relat Cancer 8(3):219–225

Ma S, Chan KW, Hu L, Lee TK, Wo JY, Ng IO, Zheng BJ, Guan XY (2007) Identification and characterization of tumorigenic liver cancer stem/progenitor cells. Gastroenterology 132(7):2542–2556

Ma S, Lee TK, Zheng BJ, Chan KW, Guan XY (2008) CD133$^+$ HCC cancer stem cells confer chemoresistance by preferential expression of the Akt/PKB survival pathway. Oncogene 27(12):1749–1758

Pang RW, Poon RT (2007) From molecular biology to targeted therapies for hepatocellular carcinoma: the future is now. Oncology 72(Suppl 1):30–44

Pardee AD, Butterfield LH (2012) Immunotherapy of hepatocellular carcinoma: Unique challenges and clinical opportunities. Oncoimmunology 1(1):48–55

Rinaldi M, Iurescia S, Fioretti D, Ponzetto A, Carloni G (2009) Strategies for successful vaccination against hepatocellular carcinoma. Int J Immunopathol Pharmacol 22(2):269–277

Shimizu S, Takehara T, Hikita H, Kodama T, Tsunematsu H, Miyagi T, Hosui A, Ishida H, Tatsumi T, Kanto T, Hiramatsu N, Fujita N, Yoshimori T, Hayashi N (2012) Inhibition of autophagy potentiates the antitumor effect of the multikinase inhibitor sorafenib in hepatocellular carcinoma. Int J Cancer 131(3):548–557

Sun JC, Pan K, Chen MS, Wang QJ, Wang H, Ma HQ, Li YQ, Liang XT, Li JJ, Zhao JJ, Chen YB, Pang XH, Liu WL, Cao Y, Guan XY, Lian QZ, Xia JC (2010) Dendritic cells-mediated CTLs targeting hepatocellular carcinoma stem cells. Cancer Biol Ther 10(4):368–375

Suo A, Zhang M, Yao Y, Zhang L, Huang C, Nan K, Zhang W (2012) Proteome analysis of the effects of sorafenib on human hepatocellular carcinoma cell line HepG2. Med Oncol 29(3):1827–1836

Tian G, Liu J, Zhou JS, Chen W (2009) Multiple hepatic arterial injections of recombinant adenovirus p53 and 5-fluorouracil after transcatheter arterial chemoembolization for unresectable hepatocellular carcinoma: a pilot phase II trial. Anticancer Drugs 20(5):389–395

Tomokuni A, Eguchi H, Tomimaru Y, Wada H, Kawamoto K, Kobayashi S, Marubashi S, Tanemura M, Nagano H, Mori M, Doki Y (2011) miR-146a suppresses the sensitivity to interferon-α in hepatocellular carcinoma cells. Biochem Biophys Res Commun 414(4): 675–680

Wang Y, Lee AT, Ma JZ, Wang J, Ren J, Yang Y, Tantoso E, Li KB, Ooi LL, Tan P, Lee CG (2008) Profiling microRNA expression in hepatocellular carcinoma reveals microRNA-224 up-regulation and apoptosis inhibitor-5 as a microRNA-224-specific target. J Biol Chem 283(19):13205–13215

Wang B, Hsu SH, Majumder S, Kutay H, Huang W, Jacob ST, Ghoshal K (2010) TGFβ mediated upregulation of hepatic miR-181b promotes hepatocarcinogenesis by targeting TIMP3. Oncogene 29(12):1787–1797

Wang Y, Toh HC, Chow P, Chung AY, Meyers DJ, Cole PA, Ooi LL, Lee CG (2012a) MicroRNA-224 is up-regulated in hepatocellular carcinoma through epigenetic mechanisms. FASEB J 26(7):3032–3041

Wang W, Zhao LJ, Tan YX, Ren H, Qi ZT (2012b) MiR-138 induces cell cycle arrest by targeting cyclin D3 in hepatocellular carcinoma. Carcinogenesis 33(5):1113–1120

Wei W, Hu Z, Fu H, Tie Y, Zhang H, Wu Y, Zheng X (2012) MicroRNA-1 and microRNA-499 downregulate the expression of the ets1 proto-oncogene in HepG2 cells. Oncol Rep 28(2):701–706

Wilhelm SM, Carter C, Tang L, Wilkie D, McNabola A, Rong H, Chen C, Zhang X, Vincent P, McHugh M, Cao Y, Shujath J, Gawlak S, Eveleigh D, Rowley B, Liu L, Adnane L, Lynch M, Auclair D, Taylor I, Gedrich R, Voznesensky A, Riedl B, Post LE, Bollag G, Trail PA (2004) BAY 43-9006 exhibits broad spectrum oral antitumor activity and targets the RAF/MEK/ERK pathway and receptor tyrosine kinases involved in tumor progression and angiogenesis. Cancer Res 64(19):7099–7109

Wilhelm S, Carter C, Lynch M, Lowinger T, Dumas J, Smith RA, Schwartz B, Simantov R, Kelley S (2006) Discovery and development of sorafenib: a multikinase inhibitor for treating cancer. Nat Rev Drug Discov 5(10):835–844

Wong QW, Ching AK, Chan AW, Choy KW, To KF, Lai PB, Wong N (2010) MiR-222 overexpression confers cell migratory advantages in hepatocellular carcinoma through enhancing AKT signaling. Clin Cancer Res 16(3):867–875

Xu Y, Liu L, Liu J, Zhang Y, Zhu J, Chen J, Liu S, Liu Z, Shi H, Shen H, Hu Z (2011a) A potentially functional polymorphism in the promoter region of miR-34b/c is associated with an increased risk for primary hepatocellular carcinoma. Int J Cancer 128(2):412–417

Xu Y, Xia F, Ma L, Shan J, Shen J, Yang Z, Liu J, Cui Y, Bian X, Bie P, Qian C (2011b) MicroRNA-122 sensitizes HCC cancer cells to adriamycin and vincristine through modulating expression of MDR and inducing cell cycle arrest. Cancer Lett 310(2):160–169

Yang Z, Ho DW, Ng MN, Lau NCK, Yu WC, Ngai P, Chu PWK, Lam CT, Poon RTP, Fan ST (2008) Significance of CD90$^+$ cancer stem cells in human liver cancer. Cancer Cell 13:153–166

Yang F, Yin Y, Wang F, Wang Y, Zhang L, Tang Y, Sun S (2010) miR-17-5p Promotes migration of human hepatocellular carcinoma cells through the p38 mitogen-activated protein kinase-heat shock protein 27 pathway. Hepatology 51(5):1614–1623

Yao J, Liang L, Huang S, Ding J, Tan N, Zhao Y, Yan M, Ge C, Zhang Z, Chen T, Wan D, Yao M, Li J, Gu J, He X (2010) MicroRNA-30d promotes tumor invasion and metastasis by targeting Galphai2 in hepatocellular carcinoma. Hepatology 51(3):846–856

Ying Q, Liang L, Guo W, Zha R, Tian Q, Huang S, Yao J, Ding J, Bao M, Ge C, Yao M, Li J, He X (2011) Hypoxia-inducible microRNA-210 augments the metastatic potential of tumor cells by targeting vacuole membrane protein 1 in hepatocellular carcinoma. Hepatology 54(6):2064–2075

Yuan K, Lian Z, Sun B, Clayton MM, Ng IO, Feitelson MA (2012) Role of miR-148a in hepatitis B associated hepatocellular carcinoma. PLoS One 7(4):e35331

Zhang X, Liu S, Hu T, Liu S, He Y, Sun S (2009) Up-regulated microRNA-143 transcribed by nuclear factor kappa B enhances hepatocarcinoma metastasis by repressing fibronectin expression. Hepatology 50(2):490–499

Zhang JF, He ML, Fu WM, Wang H, Chen LZ, Zhu X, Chen Y, Xie D, Lai P, Chen G, Lu G, Lin MC, Kung HF (2011) Primate-specific microRNA-637 inhibits tumorigenesis in hepatocellular carcinoma by disrupting signal transducer and activator of transcription 3 signaling. Hepatology 54(6):2137–2148

Zheng Y, Yin L, Chen H, Yang S, Pan C, Lu S, Miao M, Jiao B (2012) miR-376a suppresses proliferation and induces apoptosis in hepatocellular carcinoma. FEBS Lett 586(16):2396–2403

Chapter 8
Conclusions

HCC is a complex disease caused by a variety of risk factors. Molecular mechanisms that contribute to the progression of cancer from an acute or a chronic form to a metastatic stage are not completely understood. The lack of accurate molecular markers for HCC diagnosis and treatment assessment has posed a major challenge in health care. As discussed herein, the expression of a large number of genes, proteins, and other molecules belonging to diverse cellular processes and pathways are altered in HCC. Therefore, no single test or set of tests is sufficient to provide an accurate assessment on hepatic tumor burden for every clinical situation. An appropriate strategy to target human HCC is to use combination therapy that attacks multiple different steps and pathways that are involved in the disease process. Molecular profiling of genes, proteins and other molecules aimed at deciphering the genotype-phenotype relationship may be helpful for developing new therapies and 'personalized medicine' for liver cancer.

Because of their involvement in diverse cellular events, miRNAs are widely studied in cancer research, and more than a hundred clinical trials are currently underway (http://clinicaltrial.gov). In addition, there is a surge in the filing of patent applications worldwide for microRNAs in cancer therapeutics (McLeod et al. 2011). HCC is a multifactorial disease, where the expression of a large number of genes, proteins, and other molecules from diverse cellular processes and pathways are altered. Therefore, the use of combination therapy that targets multiple different pathways, rather than a single test or a set of tests, might be an appropriate, albeit complex strategy to combat human HCC. MicroRNAs are ideal for this scenario as they are capable of targeting many mRNAs simultaneously. Success of this multi-pronged approach has been demonstrated by the positive therapeutic outcomes achieved with Sorafenib that can inhibit receptor tyrosine kinases of multiple signaling cascades, as well as evidence that a single molecule miR-26a can significantly reduce HCC with minimal toxicity (Kota et al. 2009). The main challenge for successful translation of these strategies into the clinical arena is to harness the full potential of microRNAs for therapeutic development.

© The Author(s) 2014
R.N. Aravalli, C.J. Steer, *Hepatocellular Carcinoma*, SpringerBriefs
in Cancer Research, DOI 10.1007/978-3-319-09414-4_8

Novel approaches to enhance the stability of microRNA modulators *in vivo*, and methods to deliver them efficiently to the target tissues are urgently needed to overcome this challenge. The next decade will undoubtedly witness the development of exciting, novel and effective approaches to treating this horrific disease.

References

Kota J, Chivukula RR, O'Donnell KA, Wentzel EA, Montgomery CL, Hwang HW, Chang TC, Vivekanandan P, Torbenson M, Clark KR, Mendell JR, Mendell JT (2009) Therapeutic microRNA delivery suppresses tumorigenesis in a murine liver cancer model. Cell 137(6):1005–1017

McLeod BW, Hayman ML, Purcell AL, Marcus JS, Veitenheimer E (2011) The 'real world' utility of miRNA patents: lessons learned from expressed sequence tags. Nat Biotechnol 29(2): 129–133

Index

A
Aberrant expression, 33, 39, 40
Adenomatous polyposis coli (APC), 34
Aflatoxin B1 (AFB1), 3, 35
Alanine aminotransferase (ALT), 52
Alcohol, 3, 5, 10, 23, 24, 34, 38
Alcohol dehydrogenase 1A1 (ALDH), 23
American Association for the Study of Liver Diseases (AASLD), 10
Anatomic resection (AR), 11
Animal model, 23, 36, 47–50, 52

C
Cancer-associated fibroblasts (CAFs), 16
Cancer stem cells (CSCs), 22–24, 53, 59
Carbonic anhydrase 9 (CA9), 19
Choline-deficient L-amino acid defined (CDAA) diet, 56
Cirrhosis, 1, 4, 5, 7, 10, 11, 15, 24, 25, 34, 38, 48–49
Clinical trial, 9, 52–55, 58, 65
C-X-C chemokine receptors (CXCRs), 17
Cyclin-dependent kinases (CDKs), 35, 36

D
Demographics, 4–5
Dendritic cells (DCs), 16–17, 53
Diagnosis, 1, 3, 7–11, 50, 56
Drug treatment, 9

E
Endothelial cells, 17–18
Epidermal growth factor (EGF), 16, 17
Epithelial-mesenchymal transition (EMT), 19, 21
Extracellular matrix (ECM), 16–19, 56, 57
Extracellular-regulated kinase (ERK) pathways, 37, 38, 51–53

F
Fibroblast growth factor (FGF), 16, 18, 38, 52

G
Gender difference, 5
Genetic heterogeneity, 10–11
Glutathione S-transferase (GST), 35

H
Hand-foot skin reactions (HFSR), 52
Heat shock proteins (HSPs), 38
Hepatic stellate cells (HSCs), 15–17, 25, 37
Hepatitis virus, 4
Hepatocellular carcinoma (HCC), 1, 3, 7, 15, 33, 47, 51, 65
    cellular signaling pathways, 3, 4
    cirrhosis, 5
    clinical features, 7
    genetic heterogeneity, 10–11

© The Author(s) 2014
R.N. Aravalli, C.J. Steer, *Hepatocellular Carcinoma*, SpringerBriefs in Cancer Research, DOI 10.1007/978-3-319-09414-4

Hepatocellular carcinoma (HCC) (*cont.*)
  heavy alcohol consumption, 5
  interferon therapy, 9
  medical care expenses, 1
  metabolic syndrome, 3
  miRNA therapeutics, 55–59
  pigs, 48–49
  primates, 49
  rabbits, 48
  rats, 48
  signaling pathways, 33–38
  small molecule-based therapeutics, 51–55
  therapy, 25, 34–36, 51–60
  treatment, 1, 5, 7–11, 33, 40, 47, 48, 51, 57,
    59, 60, 65
  tumor initiation and progression, 16
  woodchuck hepatitis virus (WHV), 49
Hepatocytes, 15, 19, 21, 24, 25, 35, 36
HSPs. *See* Heat shock proteins (HSPs)
Human progenitor cells (HPCs), 23
Hypoxia, 19–21, 56, 58

I
Inflammation, 15, 17, 24–25, 34, 38
Insulin-like growth factor 1 receptor (IGF-1R),
  33, 58
Intervention, 7–10, 48, 56–57

J
Janus kinase (JAK) pathway, 37–38, 52

K
Kupffer cells, 15–17, 22, 24–25

L
Liver cancer, 1, 3, 4, 21, 24, 33–38, 41, 47–50,
  53, 57, 65
Liver sinusoidal endothelial cells, 24

M
Mammalian target of rapamycin (mTOR), 52,
  58
Matrix metalloproteases (MMPs), 16, 17, 25
MicroRNAs (miRNAs), 33, 39–41, 55–59,
  65–66
Mitogen-activated protein kinase (MAPK)
  pathway, 36–38, 52, 56

Model for End-stage Liver Disease (MELD),
  11–12
Molecular marker, 51–54, 65
Molecular mechanisms, 3, 33–41, 47, 50
Multi-pronged approach, 65
Myeloid cell leukemia-1 protein (Mcl-1), 38

N
Natural killer (NK) cells, 15, 25
Nonalcoholic steatohepatitis (NASH), 5
Non-AR (NAR), 11

O
Obesity, 3, 24, 37–38
Orthotopic liver transplantation (OLT), 1
Oxidative stress, 21–22, 25, 30, 38

P
Phosphoinositide 3-kinase (PI3K) pathway, 19,
  20, 23
Pig, 48–49
Platelet derived growth factor (PDGF), 16–18,
  21, 37
p53 pathway, 33, 35, 36, 38, 56
Primate, 49
Programmed death ligand 1 (PD-L1), 16

R
Rabbit, 48
Ras pathway, 36
Rb pathway, 33, 35–36
Risk factor, 3–5, 10, 11, 15, 33, 65
Rodent, 35, 36, 47

S
Secreted protein acidic rich in cysteine
  (SPARC), 16
Signal transducer and activator of transcription
  (STAT) pathway, 24, 37–38, 57
Signal transduction, 39, 47, 53
Sorafenib, 9, 41, 51–55, 65
Stem cell, 59–60
Superoxide dismutase (SOD), 22
Surveillance, epidemiology and end results
  (SEER), 1

T
T cells, 16–18, 25
Therapeutic tool, 50

Therapy, 1, 8, 9, 19–22, 25, 41, 49, 50, 53, 54, 57–59, 65
Tumor-associated macrophages (TAMs), 16–17
Tumor microenvironment, 16–19
Tumor necrosis factor-α (TNF-α), 16–18, 24, 25, 37–38

V
Vascular endothelial growth factor (VEGF), 16–18, 20, 38

W
Wnt/β-catenin signaling pathway, 34–35, 54
Woodchuck, 49